Biological Energy Resources

Biological Energy Resources

Malcolm Slesser
Director
Energy Studies Unit
Strathclyde University

and

Chris Lewis
Senior Research Fellow
Strathclyde University

London
E & F N Spon Ltd

A Halsted Press Book, John Wiley & Sons, New York

First published 1979
by E. & F.N. Spon Ltd
11 New Fetter Lane, London EC4P 4EE

Printed in Great Britain by
Richard Clay (The Chaucer Press) Ltd
Bungay, Suffolk

ISBN 0 419 11340 1

Distributed in the U.S.A. by Halsted Press,
a Division of John Wiley & Sons, Inc., New York

Library of Congress Cataloging in Publication Data

Slesser, Malcolm.
Biological energy resources.

Bibliography: p.
Includes index.
1. Biomass energy. I. Lewis, Chris, joint
author. II. Title.
TP360.S53 662′.6 79-10255
ISBN 0-470-26729-1

Contents

Preface

As it becomes increasingly clear that those wonderfully convenient energy sources, oil and gas, cannot last the world for ever, man has begun to turn his mind to alternatives. Far ahead in the game is nuclear energy, a proven technology, but one which is rapidly becoming more and more capital-intensive. Moreover there is a very substantial amount of world public opinion opposed to it because of the inherent risks involved. Another lobby has now grown up, that devoted to soft energy paths, particularly to harnessing energy from the sun. This book is about one of the ways in which the sun furnishes energy, namely by photosynthesis to produce what are now called the biological energy resources or 'biofuels'.

While there are no shortage of ideas and of proponents, there *is* a shortage of solid facts upon which to judge whether bioenergy resources can be economic, and whether they can indeed provide an alternative path to the concentrated, centralized energy systems that typify, for example, nuclear energy. This book sets out to provide a framework by which biological energy resources may be judged economically, and to set the frame within which we can look for improvements in their performance.

The book does not claim to espouse biological energy resources as the correct or ideal path to future energy supplies. In fact we have tried to keep our value judgements out of the analysis. We have shown that some propositions are hardly feasible, but that, given the right circumstances, many others are. It is the nature of these 'circumstances' that constitutes the most important part of this book.

We hope that it will help researchers, innovators and decision makers to see more clearly through the trees of knowledge to the 'biofuels' which flow beyond.

M Slesser
C. Lewis

October 1978

Acknowledgements

This book was made possible through the original endeavours of many pioneers in bioenergy research. In particular the authors are grateful for discussions, held over a number of years, with several workers covering a number of disciplines. We should like to cite for special mention:

Professor David Hall, King's College, University of London, UK
Professor William Oswald, University of California, Berkeley, USA
Dr John Benemann, University of California, Berkeley, USA
Dr Joel Goldman, Woods Hole Oceanographic Institution, Massachusetts, USA
Dr Cesar Marchetti, International Institute for Applied Systems Analysis, Austria
Dr Roger Holdom, University of Manchester Institute of Science and Technology, UK

Special thanks are also due to Miss Elizabeth Baynes, Miss Rosalind Baynes and Miss Catharine Thomas for their speedy and accurate typing.

Finally, one of us (C.L.) would especially like to extend his gratitude to his wife Anne, for her patience and understanding. We were married halfway through Chapter 4, and are still married at the completion of the book.

Chapter 1

Introduction

1.1 Bioenergy! Why it's worthwhile

For the greater part of the history of mankind, solar energy has been at the root of the driving force of his economy. This solar energy appeared to him largely in the form of wood, renewable through the process of growing a tree with its attendant photosynthesis. But there were also other forms of solar energy such as the use of the direct sun to evaporate sea water to make salt. It was the discovery of easily winnable coal that created the Industrial Revolution in England, which spread all over the world, so that today we think of a normal country as being industrialized and highly energy-intensive, and in which its inhabitants do not achieve their objectives through the sweat of their brows, but through the application of energy by the medium of fuels. In such societies solar energy has taken a back seat. Fuels, first coal, then oil, and now uranium, became steadily cheaper in relative terms as time passed. These fuels provided intensive sources of heat and enabled us to build things like steel mills and motor cars at relatively low capital costs. Economic textbooks were written in such a way that there appeared to be but one path to development, a path using fossil or perhaps fissile energy to fuel capital structures designed only to accept intensive energy sources. This route seemed quite ineluctable for developed and developing countries alike, and there was in no one's mind the possibility of choosing another path. To have suggested in 1970 that solar energy might be a means to a national development would have been to open oneself to ridicule!

In 1973 the Organisation of Petroleum Exporting Countries (OPEC), unilaterally raised the price of its commodity – oil, by 300 per cent and sent a shock wave running through the world's industrialized communities. Quite suddenly and understandably a lot of people, from politicians and policy makers to the man in the street, turned eyes back towards the sun, and asked themselves the question: was there in that solar radiation an alternative to fossil and fissile fuels? Naturally, those who were environmentally or ecologically minded, very quickly put forward arguments for the use of solar energy often backed up, one regrets to say, with very optimistic figures.

Solar energy has many roles to play; it can operate through the medium of photosynthesis, through action on the weather, which in turn can affect waves and wind. It can create energy jumps in semiconductors, it can be absorbed as a warming radiation. This book is about one only of these possible applications, it is about the use of solar energy for what are now called bioenergetic systems – that is to say, systems often of great biochemical complexity, which use the sun's radiation as the driving force for change. The book seeks to explain how these mechanisms work and whether they have a practical field of application as energy suppliers to society at one or another level.

If the answers are not entirely clear, the means of analysing their potential is now at hand and the need for research and technological development obvious. The great questions, such as: can society adapt itself to a bioenergetic means of sustaining its economy? are by no means clear. Nor yet are the questions which are often asked: can bioenergetic routes provide a better means for the development of less developed countries? It is hoped that the ideas and analyses set forth in this book will help those who are interested in gaining deeper insight into these matters to come to sound conclusions.

At the outset it may be useful briefly to review why it is worth having a serious look at biological paths to the production of energy and energy-containing materials. By 1978, it had become quite apparent that, given the growth rate of population; given the objectives of economic growth; given the stock of accessible reserves of coal, oil and gas, it was going to be impossible to provide as much energy to the world economy in the later years of this century as was going to be demanded, and that if there was not to be an energy shortfall, nuclear energy would have to provide an interim solution. Moreover, if some other alternative was not developed, nuclear energy would have to represent the final and total solution. Now, not everybody is enthusiastic about a nuclear predicated world, for reasons which are associated with, on the one hand, possible dangers of radioactivity and, on the other hand, the political institutions that might develop in a world dominated by nuclear fission or fusion. Bioenergy technologies are examples of what Amory Lovins [1] calls 'soft energy paths'. But even a cursory study of these systems demonstrates that while they may be soft, they are far from being technologically simple or easy to operate. They carry with them their own intrinsic complexities, which must be understood and mastered in just the same way that a nuclear reactor may not be safely run, except by people who understand the underlying physics and the means of control. Yet another driving force in the direction of the bioenergies is the feeling that they are more consistent with a world in which we appeared to be running into an excess of people with insufficient work for them to do. Moreover, traditional methods of economic development require quite massive capital investments, whereas for the biotechnologies there are only three basic parameters involved: land, know-how, and solar energy. In relative terms, bio-enthusiasts argue that the capital investments are smaller and the inputs essentially zero. This is undoubtedly a somewhat Utopian view of the biotechnologies, and we hope in the course of this book to reveal more clearly what are the issues at stake in their development and use. Finally even in the world of conventional fuels there is an urgent call for conservation, and the bioenergy technologies represent an excellent way of implementing this conservation.

1.2 Photosynthesis and fuel production – In a nutshell

The annual quantity of solar radiation incident at the earth's surface is approximately 3×10^6 EJ [2], equivalent to about 75 times the present total proven world fossil fuel energy reserves. Of this vast energy influx, at maximum 60 per cent is available for absorption by marine and terrestrial plant life, the remainder being reflected by deserts, ocean surfaces, snow, ice, etc. The percentage of the potentially available energy which is in the photosynthetically active radiation (PAR) region, within spectral wavelengths from 400 mm to 700 mm, amounts to 43 per cent. In the case of land plants the optimally arranged leaf canopy can absorb 80 per cent PAR, the efficiency of the absorbed energy's conversion in the photosynthetic reactions is 23 per cent, and, after allowing for respirational losses, net photosynthesis is further reduced to 66.7 per cent. This represents an overall efficiency of 5.3 per cent which is in fact virtually a presently attainable maximum [3]. In effect, the average efficiency usually attained by most plants under field conditions seldom exceeds 1–2 per cent (see Section 3.1), although advances made via plant biochemical, physiological, and genetical research promise to improve this situation in the future. Increased utilization of plants such as sugar cane, sorghum, and maize, which contain the more efficient 4-carbon photosynthetic pathway

(Section 2.2) should result in the mean plant photosynthetic energy capture efficiency being increased on an overall global scale. Low efficiency automatically means extensive land area requirements if biomass is ever to provide a significant proportion of man's energy needs during the coming decades.

There are three non-mutually exclusive approaches which can be adopted to obtain suitable organic raw material starting points for energy production from biomass: the purposeful cultivation of specific 'energy crops'; the harvesting of natural vegetation; and the use of agricultural and other wastes, both of plant and animal origin. The subsequent routes by which this biomass may be transformed into high energy products are many and various, and include fermentation, anaerobic digestion, biophotolysis, combustion, hydrogenation, partial oxidation, and pyrolysis.

Perhaps the most frequently produced fuels are methane, methanol, and ethanol; although, depending on the initial substrate and conversion technology employed, solid fuels, hydrogen, low-energy gases, and long-chain hydrocarbons can also be formed. A further conceivable option is the generation of electricity or heat directly by combustion of biomass with a low water content, especially matter rich in lignin and cellulose, such as wood. Biomass sources with a high water content are usually more advantageously utilized as fermentation substrates so as not to incur an energy penalty for the otherwise required dehydration process.

Three of the most relevant factors to any energy consideration of a particular process are its location, its desired level of intensity, and the extent of any pre-treatment required prior to the occurrence of the conversion operation itself. These various parameters, along with net energy analyses of selected processes, will be dealt with in the appropriate ensuing chapters.

1.3 Biological energy resources in the past

The concept of a biological energy supply for man is by no means a new one of course, since the combustion of wood and other organic matter as a principal fuel source has been practised for millennia, and still is in much of the developing world. Neither is the phenomenon of an energy crisis a prerogative of the 1970s! During the sixteenth and more than half of the seventeenth centuries the consumption of Britain's basic fuel supply, wood relentlessly increased as the population expanded and industry devoured more, while large areas of woodland were simultaneously cleared to provide extra land for cultivation [4]. The inevitable outcome was a serious shortfall in supply, prices soared, and it was only after much readjustment over a prolonged period that gradual replacement by coal came into effect. In the seventeenth and eighteenth centuries particularly, much forest area was cleared in the Scottish Highlands to produce charcoal for the smelting of iron [5]. Indeed, it has been estimated that in 1850 wood met 91 per cent of US fuel requirements [6], and constituted the country's prime energy source until about 1875 [7]. On a global basis, consumption of fuelwood is judged to have risen from around 500 million m^3 per annum just prior to the First World War to 1070 million m^3 in 1950, at which level annual consumption has approximately remained constant, with a consequent decrease on a *per capita* basis. One estimate is that wood provided 40 per cent of world energy supplies before the Industrial Revolution, but now accounts for only 2 per cent [8], while another quoted figure is that fuelwood use, both for energy and other forest products, constitutes the equivalent of about 15 per cent of the world's annual commercial fuel consumption [9]. Whatever the precise statistic, which there is no certain way of calculating, the fact remains that over 90 per cent of the Third World's population depends upon firewood as its main fuel source, when often the rise in numbers of new trees is insufficient to meet the demands of the increase in people [6]. As examples, in Tanzania, Gambia, and Thailand the utilization of woodfuel *per capita*/year is 1.8 t, 1.2 t, and 1.1 t respectively; the woodfuel proportion of total timber consumption is 96 per cent, 94 per cent, and 76 per cent; and the population percentage using woodfuel is 99, 99, and 97 [9].

Alcohols too, have been used by mankind since the dawn of civilization, though not always as an energy source in the accepted sense. Ethanol was known to the Arabs and the Romans before the time of Christ [10], and, according to an order of the Governor of Jamaica in 1661 alcohol was being produced from the juices of sugar cane and termed 'rum' [11]. Indeed, until about 1929 virtually all US ethanol was produced via the fermentation of grains, molasses, and other materials rich in starch or sugar, and alcohols were also used extensively as fuels before competition by the presence of abundant petroleum supplies [10]. Alcohol became a popular fuel for lighting in about 1830, when it replaced malodorous fish and whale oils, but was itself substituted by the brighter kerosene flame 50 years later. In provincial France during the mid-nineteenth century wood was distilled for methanol production which, in turn, provided a heating, lighting and cooking fuel for the citizens of Paris [7].

With the advent of the internal combustion engine, fermentation alcohol rapidly made inroads into national transport sectors, with the Germans probably the pioneers in the early 1900s. Various raw materials were utilized as substrates – yams, cassava, and sweet potatoes, as well as the better known sugar cane and wood. A company was selling automobile fuel in north-eastern Brazil in the 1930s consisting of 75 per cent alcohol and 25 per cent ether, the alcohol being derived from sugar cane [12]. Both World Wars, of course, saw petroleum shortages in several countries, and particularly the Germans and French employed wood burners in many vehicles, including aeroplanes and tanks. The distillation of wood chips gave rise to alcoholic gases, some carbon monoxide, and some hydrogen, by which the vehicle continued to function, if not without incident. By 1938 some 9000 wood-burning cars were in use [13], and at the same time the European consumption of absolute alcohol as a motor fuel constituent topped 0.5 million t, with a global total of 0.57 million t. The fermentation of hydrolysed wood wastes to produce ethanol was occurring in over 20 plants outside the USA in 1941, with the dilute acid Scholler process the most favoured in Germany, Italy, and Switzerland. The German Holzminden plant was capable, in ideal operating conditions, of producing 3500 t ethanol each year from 22 000 t of wood [14], while one Finnish operation processed roughly 7000 m^3 of pine per 24-hour shift [15]. In the United States and Canada many commercial enterprises have sprung up over the years, dating as far back as 1915 when alcohol was fermented from sawdust previously hydrolysed by dilute sulphuric acid in the states of Louisiana and South Carolina. An extension of this procedure resulted in the further conversion of ethanol to ethylene in 96 per cent yield, and to butadiene in 70 per cent yield, by the USA during World War II in plants that required appreciably lower capital costs than those based on hydrocarbon conversion [16].

Since the last war, alcohol/petrol motor fuel mixtures have been continued to any great extent only in a few countries such as India, the Philippines, Argentina, and Brazil. Brazilian industry has relied for decades on cassava starch for the production of high quality ethanol, and back in 1946 over 9 million t of cassava were cultivated [17]. At the present time biomass plantations based on sugar cane and cassava combinations are planned in an effort to reduce the country's substantial expenditure on imported oil. The fuel crops will be fermented to anhydrous ethanol so as to provide 3 million m^3/annum for inclusion in motor fuel up to a level of 20 per cent. The construction of a further 193 distilleries has been authorized to accommodate the vastly increased load, so that Brazil should be self-sufficient in alcohol by 1984 at the latest [18–21]. Even in Western countries, where the major process of current industrial ethanol production employs ethylene as its raw material, alcoholic fermentation still has a place. For example, in 1962 grain fermentations accounted for 22 per cent of US industrial alcohol production, and with the escalation of petroleum prices, fermentation alcohols become economically attractive once more.

The commercial application of anaerobic methane fermentation dates back to the last century, when it was carried out merely to stabilize and humify organic wastes (i.e. in septic tanks). In 1897 a waste-disposal tank serving a leper colony in Matunga, Bombay is reported to have been equipped with gas collectors and the gas used to drive engines. The city of Fort Dodge, Iowa had, in 1954, a 250 hp gas engine driving a 175 kW generator, the fuel for which was methane produced at the local sewage treatment plant [22]. Within the UK, the world's first large-scale anaerobic digester plant to stabilize the sewage sludge produced by the then one million population of Birmingham was built in 1911, and subsequently a large gas engine generator was added to convert the digester gas into lighting and heating electricity [23]. Several of the larger sewage works operate such digesters at present, while on the small scale the rural Gobar 'biogas' plants of rural India and China have produced methane from human and livestock manure for many years.

There have been countless other examples of man utilizing natural biological phenomena as an energy source; but from where does this energy originate and in what form, and how is it made available for human exploitation? Chapters 2, 3 and 4 aim to provide some of the answers to these fundamental questions, while Chapters5, 6 and 7 are concerned with the energetic implications, the economics involved, and the present state and future prospects of biomass energy systems respectively. It should be noted that 'dollars' refer to US dollars throughout, 'pounds' are always pounds sterling, and during the period for which prices are quoted, one pound is approximately equal to just under two US dollars.

Chapter 2

Solar energy and photosynthesis

2.1 Solar energy – Down to earth

It is estimated that energy radiates from our sun at the 'astronomical' rate of some 10^{23} kW. Thermonuclear fusion reactions at the sun's central core temperature of something just below 2×10^7 °C result in the collision and coalescing of hydrogen ions, such that four protons combine to form a helium nucleus with a frequency so great that 3.6×10^{26} J of energy are released per second. Translated into units of mass this is equivalent to 6×10^8 t of helium being so formed every second, according to the simplified equation:

$$4\,H^1 \rightarrow He^4$$

At the same time are liberated particles called positrons ($\beta+$), which collide with unattached electrons within the plasma core and are obliterated, and neutrinos ($\overline{V}e$), escaping into space as unreactive entities, as well as gamma radiation (γ). The total amount of energy made available by the formation of a single helium nucleus is of the order of 5×10^{-12} J, and initially is present as the very short wave γ-radiation, with wavelengths approximating 10^{-11} m.

On their journey outwards to the sun's surface, through a mass of about 80 per cent hydrogen and 20 per cent helium nuclei together with free electrons, the units, or quanta, of radiation energy inevitably undergo numerous collisions with these particles. This series of events renders the radiation progressively less energetic and so increases the wavelengths correspondingly, with the result that in the sun's photosphere region near its surface the entire spectrum of electromagnetic radiation is found, through X-rays at 10^{-9} m to the ultra-violet, visible light, and infra-red regions, and on to low-energy microwaves and finally beyond into wavelengths of several hundred metres. The temperature of the sun's surface is between 5500 °C and 6000 °C and, at this temperature, the greatest intensity of emitted light falls in the visible spectrum at around 600 nm, coinciding with the orange zone. About 98 per cent of the total emitted energy from the photosphere is conveyed by radiation of wavelengths between 250 and 3000 nm, with as much as 50 per cent from 350 to 750 nm. This energy distribution approximates that of the so-called 'black body' (an ideal substance which would absorb all the radiation falling on it and emit nothing or, equally, emit radiation and absorb none depending on the prevailing conditions) at the appropriate temperature.

The energy of each quantum of radiation is that expressing the minimum value which can change in the confines of a system exhibiting wave-like characteristics, and the quantum itself is directly proportional to the frequency of its associated radiation. The energy of a single quantum of light radiation is termed a photon, and is equal to about 3×10^{-19} J. It is the product of the frequency of light and the very small value of Planck's constant (6.6×10^{-34} Js), expressed in the relationship $E = hv$, where E is the energy of a single quantum. Thus it follows that for quantum energy level changes to occur the radiation frequency, 'v' must be of a high magnitude. As wavelength and frequency are inversely related, then photons of shorter wavelength radiation are more energetic than those of longer wavelength radiation. Thus, within the sun, quanta of the originating γ-radiation undergo collisions with particles like protons and electrons until, near the surface, where intact atoms can exist, a collision could cause the displacement of an electron from an inner to an outer orbit. This electron then relinquishes its acquired energy in radiation of a characteristic frequency on resuming its normal condition once more, and so the complete radiation spectrum is formed and is emitted into space. In order for photosynthesis to occur on earth a similar kind of electron displacement mechanism takes place within appropriate plant pigments. The energy of a photon is absorbed at a characteristic wavelength to bring this about, and thereby a chain of photosynthetic chemical reactions is set in motion.

However, by no means all the radiation leaving the vicinity of the sun is destined to reach earth. In fact the total power intercepted by this planet is 1.8×10^{14} kW, equivalent to 5.6×10^{24} J of energy per annum; and 3×10^{24} J of this value, i.e., about 50 per cent actually arrives at the earth's surface each year, having travelled through the height of the atmosphere.

This last figure is equivalent to approximately 75 times the present total proven world fossil fuel reserves, but must be set beside its diffuse and variable nature, with a peak solar energy flux of around 1 kW/m^2 only.

The solar constant, which is defined as the extra-terrestrial irradiance normal to the solar beam just outside the earth's atmosphere, has a mean value of 1.36 kW/m^2, although changes in the pattern of global irradiance are frequent, being particularly influenced by the inclination of the earth's axis to the plane of its orbit at approximately 66.5°. At a particular location, the fluctuation in the incident solar radiation may be described as a function of latitude, season, and time of day, and is naturally a profound influence upon climate and hence biological primary productivity. As it journeys through the earth's atmosphere the solar spectrum is considerably modified, and, as stated previously, nearly 50 per cent of the total energy is removed, by both scattering and absorption. On a global average some 25 per cent of the incident solar radiation is reflected back into space by clouds, while a further 5 per cent is lost by scattering due to dust particles, water vapour, and other atmospheric molecular constituents.

Absorption is a far more selective phenomenon with respect to the wavelength of the incoming radiation. Virtually all the radiation in the ultra-violet range is absorbed by the ozone layer in the uppermost reaches of the atmosphere, where dissociation and recombination of O_3, O_2, and O occur to form a steady-state condition. Further substantial reductions in this potentially biologically harmful zone of the spectrum below 350 nm wavelength are brought about by absorption due to N_2 and N, while molecules of water vapour and carbon dioxide absorb most strongly in the near and middle infra-red respectively. Thus absorption accounts for the loss of a further 20 per cent of the original solar constant, although the visible region is relatively clear of absorption bands.

A further mean value of 5 per cent of the incident radition is reflected back from the earth's surface, but this can be a very variable quantity, with reflectivity from ice sometimes as much as 50 per cent and from still water only 2 per cent. 60 per cent of the solar energy which does reach the surface is a component of the sun's direct beam, the remainder having been scattered downwards by the atmosphere as diffuse short-wave radiation. However, these proportions are very much subject to locality, season, and the pertaining atmospheric conditions. A UK winter, for example, may have a radiation regime with only 30 per cent direct and up to 70 per cent diffuse as its constitutents. The overall picture is thus one of quite considerable variability, and factors such as urban air pollution, differing levels of water vapour content, the geographical source of the prevalent air mass, and the actual path length of the solar beam all exert considerable influence at any particular location.

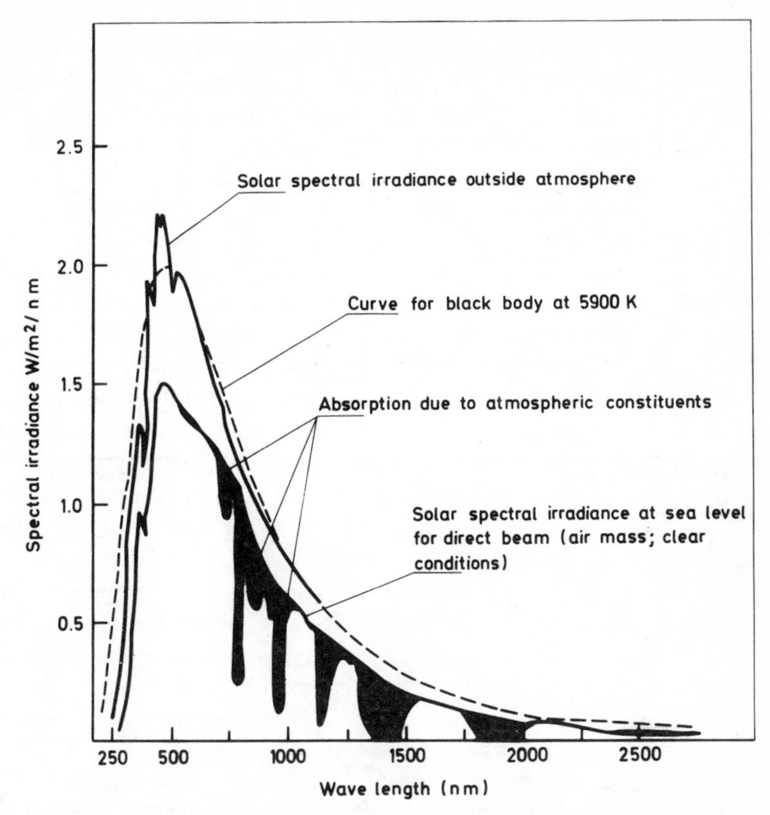

Fig. 2.1 Spectral irradiance curves for direct sunlight extra-terrestrially and at sea level (after UK-ISES [24]).

A fairly typical irradiance curve at the earth's surface for direct sunlight after passage through a clear atmosphere may be compared with the spectral irradiance prior to atmospheric entry in Figure 2.1. The radiation incident at the surface is seen to lie almost exclusively within the range 300 nm and 2500 nm (see also Figure 2.2.) due to the factors discussed above, and it can also be observed from the captions indicating those zones utilizable for various applications that the radiation spectrum from around 400 nm (violet light) to 700 nm (red light) is suitable for plant photosynthesis. An estimation of this photosynthetically amenable range, slightly extended to 300 nm, may be made from the incident global irradiation at any set location by the ratio ʅ, which equals

$$\int_{300}^{700} E_\lambda d\lambda \Big/ \int_{300}^{3000} E_\lambda d\lambda = \frac{\text{irradiance in waveband 300–700 nm}}{\text{irradiance in waveband 300–3000 nm}}$$

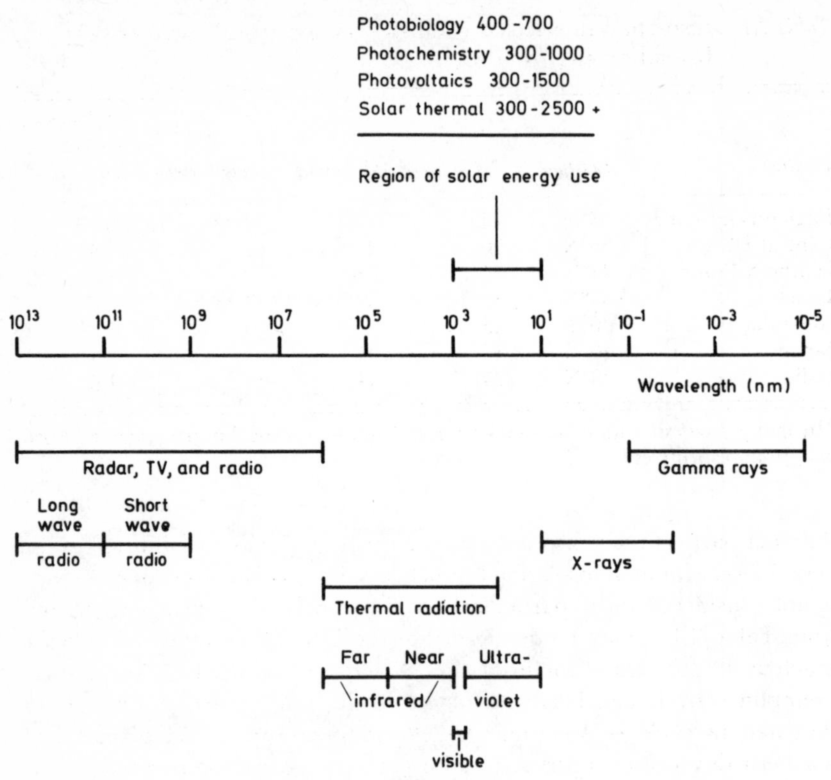

Fig. 2.2 Electromagnetic radiation spectrum (after Duffie and Beckman [25]).

where $E_\lambda \, d\lambda$ is the solar energy flux (W/m^2) in waveband of width $d\lambda$ nm. η for global radiation has been experimentally found to be 0.5 ± 0.03, with values for the direct beam only of between 0.35 and 0.45, and 0.75 for diffuse radiation in clear conditions. This value of 0.5 is virtually independent of atmospheric conditions, and is so nearly constant on a world-wide basis that regional radiation data, it has been suggested, can be confidently used to predict photosynthetic efficiencies and crop yields [26]. Thus, of the annual solar radiation incident on the earth's surface, with an energy content of 3×10^{24} J, 50 per cent or 1.5×10^{24} J is potentially photosynthetically active. 60 per cent only of this last figure is actually available for absorption by terrestrial and marine plants, since the remainder falls on oceans, deserts, ice and snow, etc. and may be reflected. The calculation for the amount of solar energy collected per annum via plant photosynthesis works out at approximately 3×10^{21} J.

As stated in the previous paragraph, the overall radiation regimes in

Table 2.1 Insolation in selected countries on horizontal surfaces [27]
Insolation (MJ/m^2 day)

Location	*Latitude*	*Midsummer*	*Midwinter*	*Annual mean*	*Midsummer/ midwinter ratio*
England and Wales	53°N	18	1.7	8.9	10.6
Central USA	36°N	26	11	19	2.4
Southern France	44°N	24	5	15	4.8
Israel	33°N	31	11	22	2.8
Australia	30°S	23	13	20	1.8
Japan	40°N	17	7	13	2.4
India	18°N	(26)*	(14)*	20	1.9

*In many parts of India the maximum and minimum insolations occur in April and June respectively.

different parts of the world may show a great degree of variability, but the actual total of annual insolation is much less diverse. North-western Europe is not considered to be particularly well endowed with solar energy, but from Table 2.1 it may be deduced that the UK, for instance, receives as much as 40 per cent of the insolation of the very favoured Middle East as exemplified by Israel. It is with regard to the relative seasonal variations that nations such as Australia, Israel and Japan are better placed than northern European countries. In addition to receiving comparatively little solar radiation in winter, when it is most needed for the space heating of buildings etc., the duration of the UK growing season for plants is also severely curtailed, with a consequent reduction in the annual crop productivities attainable in latitudes nearer the tropics.

In the absence of cloud cover and atmospheric pollution the maximum insolation is almost constant from the Equator to the Poles at 27–30.5 MJ/m^2 day for midsummer, but varies from 0 in the polar circles (latitude 66.5°), 6.1 at the mid-earth latitude of 45°, 15.1 at the tropical latitude of 23.5°, to 24.5 MJ/m^2 day at the Equator in midwinter. The overall annual total at the Equator is nearly double that at the polar circle, emphasizing once more that all these values are for clear conditions only.

The annual mean surface global irradiance on a horizontal plane as it actually occurs is presented in Figure 2.3. Values of 250 W/m^2 and above are found principally in the continental desert regions around latitudes 25° North and South, with the amount of available energy declining both towards the Equator and the Poles. The Red Sea area boasts 300 W/m^2, but cloud in the equatorial zone reduces the mean global irradiance to a very great extent. Therefore, although latitude is an important criterion when

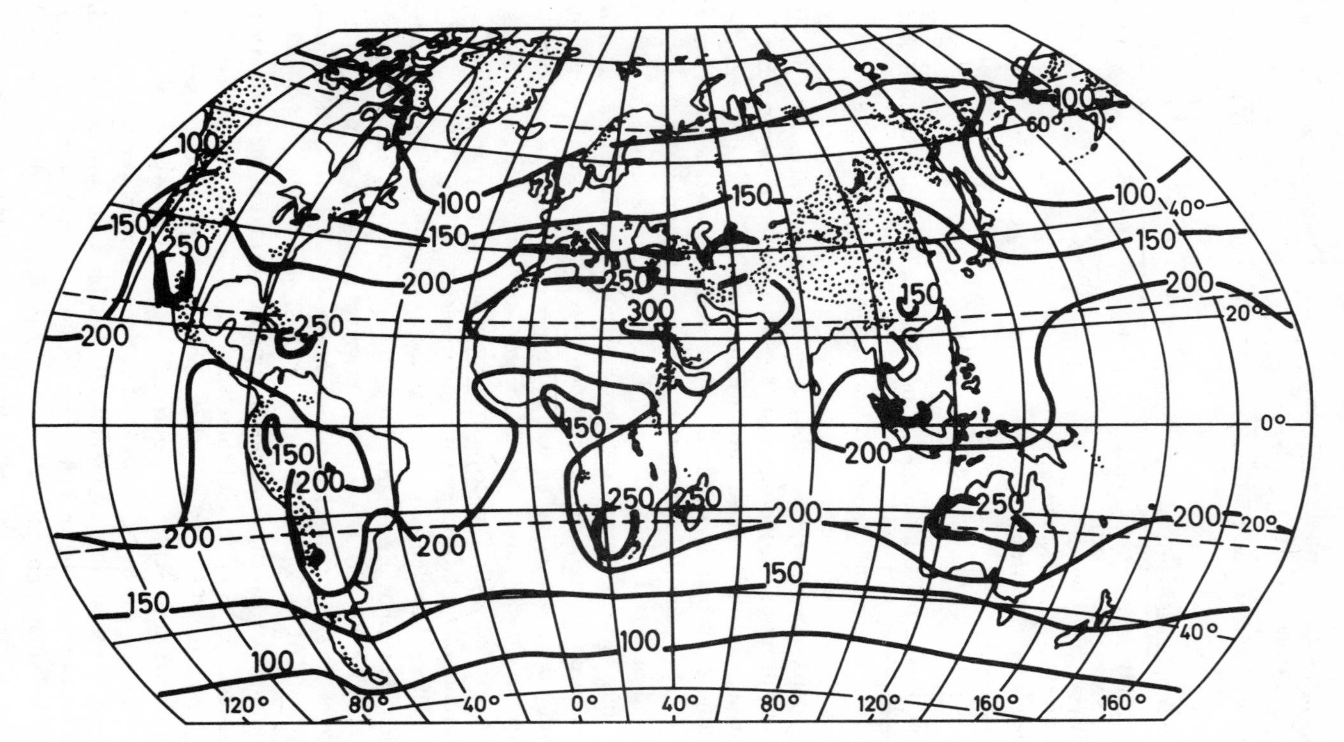

Fig. 2.3 Annual mean global irradiance (W/m^2 averaged over 24 hours) on a horizontal plane at the earth's surface (after UK-ISES [24]).

assessing insolation levels, and hence potential biological productivity, it is by no means the sole criterion. This point will be further amplified in subsequent chapters, and particularly in Section 3.1 [24, 25, 27–30].

2.2 The mechanisms of photosynthesis

Photosynthesis is the key mechanism (or, more properly, mechanisms) to which all past fossil fuel production and present food and fibre (and some energy) production owe their origins. It is the process by which a small percentage of the quanta originating from the sun, as discussed in the previous section, is collected within green plants, algae and some bacteria and converted to reduced carbon and stored energy. The problem of effective solar energy storage is still a difficult one for man to solve employing artificial devices, but plants have been doing it naturally for billions of years. It is an important concept, for the usefulness of energy capture *per se* is severely limited without an accompanying means of fairly long-term storage, so that it may be 'tapped' as and when required.

The overall photosynthetic process occurs within the chloroplasts of green plants, and the several biochemical reactions involved may be

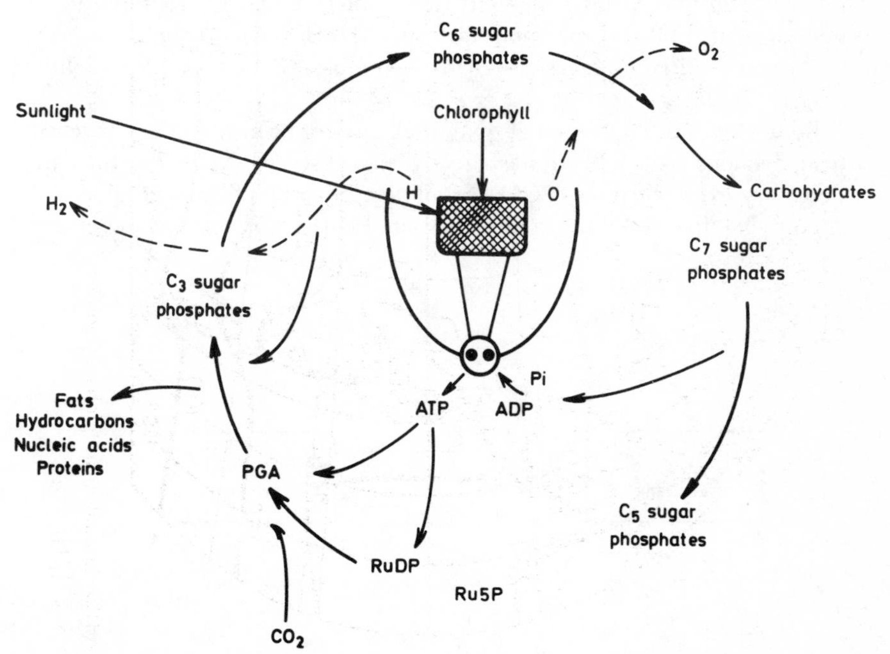

Fig. 2.4 Simplified photosynthetic reaction cycle (after Bassham [3]).

reduced to the following simplified equation:

$$CO_2 + 2H_2O + \text{light} \rightarrow (CH_2O) + H_2O + O_2 : G = 477 \times 10^3\ J$$

The light energy is absorbed by the chlorophyll and carotenoid pigments of the chloroplasts and this results in a separation of charges (Figure 2.4). A first stage reaction is set up in which electrons are removed from the water molecules to effect the evolution of gaseous oxygen at the positive charge, and with the released electrons contributing to the formation of molecules of the reducing co-factor, hydrogenated nicotinamide adenine dinucleotide phosphate (NADPH). In Figure 2.4, for convenience, the production of hydrogen is represented, which plants ordinarily do not produce in the gaseous state, but can be induced to do so under the appropriate conditions (see Section 7.2).

In addition to the formation of $NADPH_2$, energy-rich adenosine triphosphate (ATP) molecules are synthesized from the combination of adenosine diphosphate (ADP) and inorganic phosphate (Pi) in photophosphorylation reactions. Under cellular conditions ATP is capable of releasing up to 5×10^4 J of energy per mole, which, in conjunction with the reducing power of $NADPH_2$, is utilized in the subsequent enzymatic fixation of CO_2, chiefly as carbohydrate, in the absence of light. The overall reaction is summarized in Figure 2.5. In a nutshell then, each molecule of CO_2 requires three molecules of ATP and two molecules of $NADPH_2$ for its reduction to carbohydrates, initially glucose; and this synopsis will be expanded slightly later in the section.

Returning to the question of quantifiable energy considerations touched on in Sections 1.2 and 2.1, it has already been observed that 477×10^3 J are necessary to fix one mole of CO_2, entailing quite a significant energy input. Amplifying the reactions represented in Figure 2.5, the energetics are

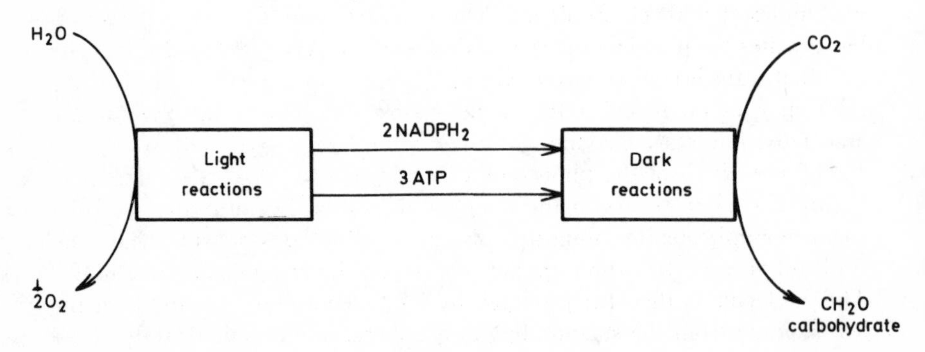

Fig. 2.5 Basic synthetic process.

shown below:

$$\text{(a) } NADPH_2 + O_2 \rightarrow 2NADP + 2H_2O.\ \Delta G = -440 \times 10^3\ J$$
$$\text{(b) } 3ATP + H_2O \rightarrow 3ADP + 3Pi.\ \Delta G = -92 \times 10^3\ J$$

Therefore, in the reduction of one CO_2 molecule to the level of glucose (requiring 477×10^3 J), reactions (a) and (b) total 532×10^3 J, thus providing an extra 55×10^3 J of energy, with over 80 per cent of the energy being derived from the reducing power of $NADPH_2$.

The efficiency of photosynthesis was discussed briefly in Section 1.2, and can be determined experimentally. A quantum of light of wavelength 680 nm (the red region) is equivalent to 176.2×10^3 J of energy, so that in order for one CO_2 molecule to be fixed as carbohydrate, 477/176.2 (or 2.7) quanta of red light need to be sequestered. In effect this means at least three quanta are required. By experimentation, 8–10 quanta of absorbed light are actually required, making CO_2 fixation around 30 per cent efficient (2.7/10 to 2.7/8) at best, with the overall plant photosynthetic efficiency rarely above 1 or 2 per cent (Sections 1.2 and 3.1), and certainly providing much capacity for improvement.

In order to describe the biochemical events of photosynthesis in more detail it is necessary firstly to further explore the 'light reactions' portion of Figure 2.5. As has been observed, the formation of ATP and the release of electrons from water are accomplished. These electrons are transported ultimately to give rise to the very reduced $NADPH_2$ molecules via a complex electron transport system. Initially the absorption of light by the chloroplastic membrane pigment P680 induces the elevation of an electron to an excited state within the so-called Photosystem II. Chlorophylls *a* and *b* are also associated with this system by forming a complex with P680. The excited electron then returns down a potential gradient through plastoquinone, cytochrome *f*, and various other pigments with the resulting formation of 'high-energy' ATP. It eventually arrives at pigment P700, which is complexed with chlorophyll *a*, where absorption of another photon of light excites the pigment and renders the electron very highly reduced on its eventual transference to ferredoxin in Photosystem I. From ferredoxin the electron goes on to effect the reduction of NADP with the appropriate associated reductase enzyme, and finally the reduction of CO_2 to sugars.

The chemistry of the photosynthetic carbon fixation cycle is shown in Figure 2.6, as it occurs in the majority of plants. Initially, the 5-carbon sugar monophosphate, ribulose-5-phosphate (Ru5P), is phosphorylated by ATP to ribulose-1, 5-diphosphate (shortened for convenience to RuDP). This molecule is then carboxylated by a CO_2 molecule, admitted via the leaf stomata from the surrounding atmosphere, in a reaction catalyzed by the enzyme RuDP carboxylase, and hydrolysed to form two 3-carbon

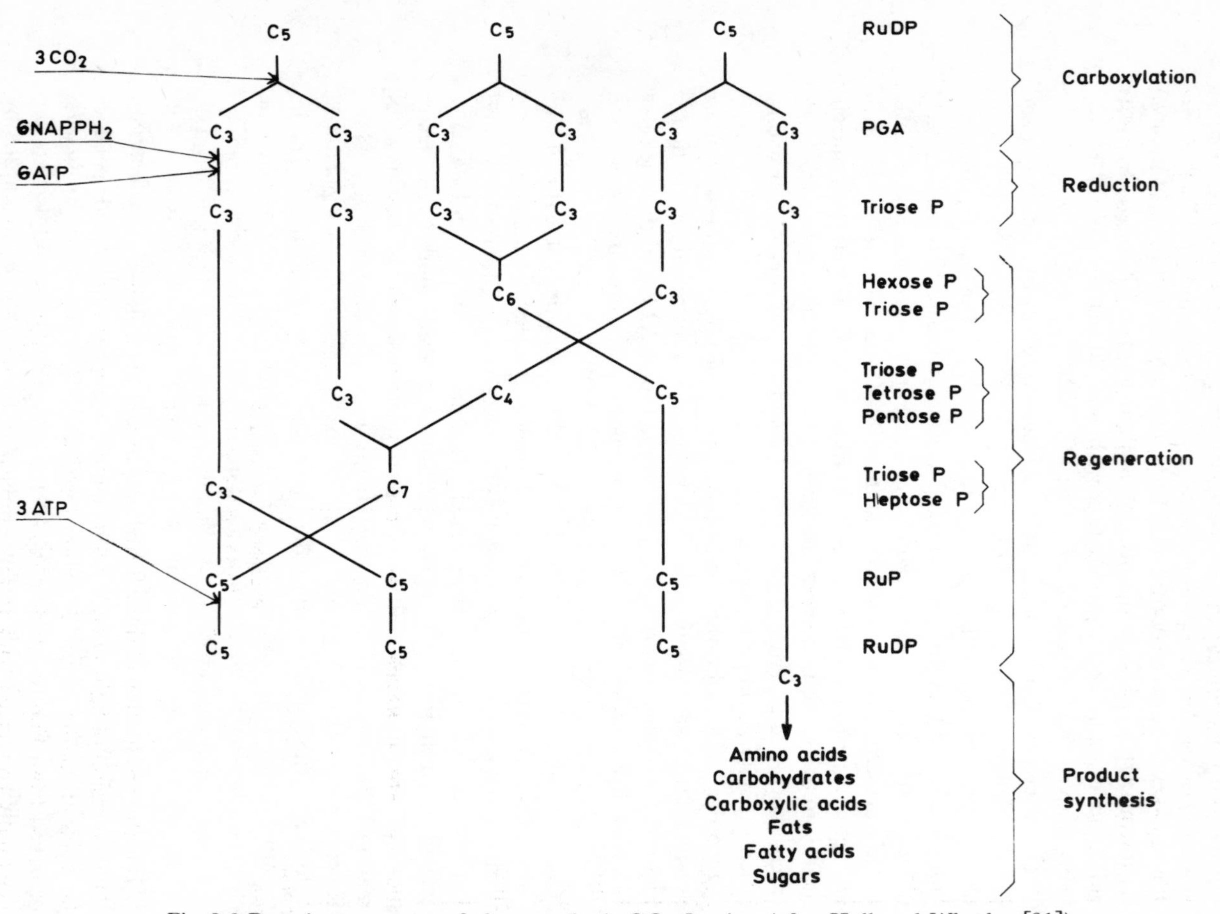

Fig. 2.6 Reaction sequence of photosynthetic CO_2 fixation (after Hall and Whatley [31]).

molecules of phosphoglyceric acid (PGA). Most of the PGA is retained in the cycle in order to regenerate RuDP, but a proportion can be transported outside the leaf chloroplasts to be utilized as an initiation point for the synthesis of various compounds, including proteins and fats. Continuing the reaction sequence, the PGA is phosphorylated by ATP and reduced with $NADPH_2$ to produce the triose phosphate, phosphoglyceraldehyde. At this stage the following principal reactions have occurred:

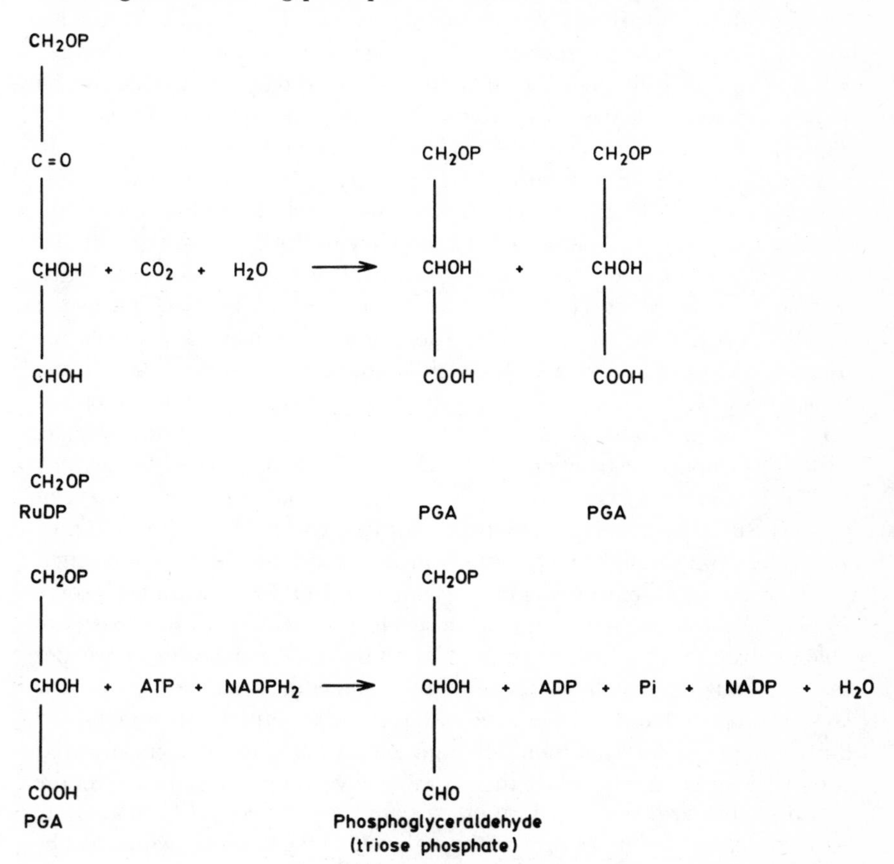

It is the production of these 3-carbon compounds which finally completes the conversion of light energy into chemical energy, and gives this particular cycle its name as the 3-carbon (or Calvin–Benson) cycle. As may be observed from Figure 2.6, five molecules of 3-carbon sugar phosphate are eventually converted back to three molecules of 5-carbon sugar phosphate (the original RuDP) in the regeneration of the initial CO_2 acceptor molecules. These reactions are inevitably complex, as depicted in the diagram. At the same time, one 3-carbon sugar phosphate remains per cycle, and is used for biosynthesis, primarily in the formation of carbohydrates. However, depending on the prevalent light, CO_2, and O_2 concentrations, differing proportions of sugars, amino acids, and fats are produced, a property which man has the clear potential to exploit advantageously [3, 30, 32].

Although the 3-carbon photosynthetic pathway is by far the most common in plants, there are species such as sugar cane, sorghum, and maize, mainly tropical, which possess an additional 4-carbon pathway besides the Calvin–Benson cycle described above. A distinguishing feature of these C_4 plants is their possession of two distinct chloroplast types, namely mesophyll and bundle sheath. The CO_2 acceptor molecules are phosphenol-pyruvate (PEP), present in the mesophyll cells and which are 3-carbon compounds. Catalyzed by the enzyme PEP carboxylase the C_1 and C_3 molecules combine to form the C_4 oxaloacetic acid molecule. This, in turn, is reduced by $NADPH_2$ to malic acid which is conveyed to the bundle sheath cells for decarboxylation, thus producing a C_3 molecule pyruvic acid, and releasing CO_2 which is now fixed once more, but in the 3-carbon cycle. Meanwhile, the pyruvic acid, on its return to the mesophyll cells, is phosphorylated by ATP to regenerate the CO_2 acceptor molecule, PEP. The important reactions of the C_4 cycle are summarized below.

PEP is more reactive than RuDP with CO_2, and furthermore oxygen inhibits RuDP carboxylase activity but not PEP carboxylase activity. Thus, when the proportions of CO_2 and O_2 in the atmosphere are low and high respectively the C_4 pathway is more efficient in CO_2 fixation. Indeed, some C_4 plants, out of the hundreds of such species which are known to exist, are able to absorb CO_2 when the concentration is down to one or two parts per million. It is principally because of this capacity to utilize CO_2 at low concentrations that 4-carbon species have a competitive advantage over 3-carbon plants in conditions of high insolation and temperature and restricted water supply; hence their comparative abundance in many of the world's more arid regions. Two other significant aspects of C_4 plants are their low rates of photorespiration and glycollate metabolism, which will be discussed shortly [33].

Before doing so, a third system of fixing CO_2 at night should be mentioned. This mechanism was originally discovered in the Crassulaceae family, thus the name Crassulacean Acid Metabolism (CAM), the distinguishing feature of which is its dramatic reduction in water loss. This renders the CAM plants suitable for colonizing desert habitats, where they are generally found. Their unique property is due to the closure of leaf stomata while radiation and temperature are great, but they are invariably slow-growing

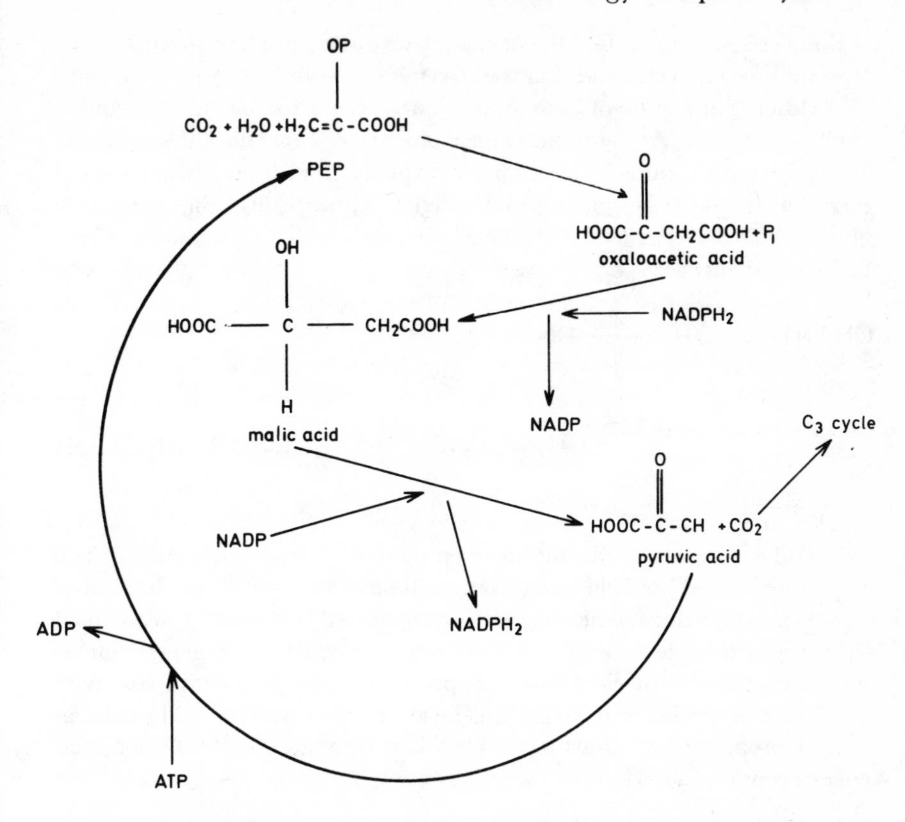

species with limited potential as providers of energy. Perhaps the most familiar example is the pineapple, which is typical of the group in having large, fleshy organs, since the system necessitates a good acid-storage capacity.

In order to survive and grow, plants need to respire aerobically and utilize oxygen from the atmosphere to oxidize mainly carbohydrates to CO_2 and water. The difference between the CO_2 assimilated in photosynthesis and that lost in respiration then becomes what is, in effect, net photosynthesis. In addition to the common dark respiration occurring in the mitochondria, plants also evolve CO_2 in a light-stimulated process known as photorespiration. Unlike mitochondrial respiration, photorespiration apparently provides no energy in a form usable by the plant cell, and thus only succeeds in decreasing productivity. Its benefical effects, if any, have yet to be fully elucidated. As stated previously, photorespiration is much greater in the inefficient C_3 species, with the rate sometimes approaching 50 per cent of that of net photosynthesis.

It now seems certain that the primary substrate of photorespiration is the C_2 acid, glycollic acid, and that the varying rates of photorespiration coincide closely with those of both glycollic acid synthesis and its subsequent oxidation to CO_2. As yet not all of the biochemical reactions necessary for glycollic acid synthesis have been explained to the satisfaction of everyone in the field, but when formed it appears that the compound liberates 50 per cent of the previously assimilated CO_2 by means of the following sequence:

$$\underset{\text{glycollic acid}}{CH_2OH - COOH} \xrightarrow[\text{oxidase}]{\text{glycollate}} \underset{\text{glyoxylate}}{CHO\text{—}COOH}$$

$$\rightarrow \underset{\text{glycine}}{CH_2NH_2COOH} \rightarrow \underset{\text{serine}}{CH_2\text{—}OHCH\text{—}NH_2COOH}$$

CO_2 is thought to arise during the step involving the condensation of two glycine molecules to yield one of serine, though the overall mechanism of CO_2 loss is probably somewhat more complicated that this. What is clear, however, is that the blocking of photorespiration, either by genetic mutation or chemical inhibition, should appreciably enhance plant productivity in numerous species, and would thus be extremely beneficial in the cultivation of proposed fuel crops to be exploited by man specifically for their energy content [30, 34–39].

Chapter 3

Biomass energy sources

3.1 Plant productivity

As mentioned briefly in Section 1.2, suitable organic materials for bioconversion are various wastes (dealt with in the following section), natural vegetation, and specific fuel crops grown for their energy content. Thus photosynthetic efficiency and crop productivity are important considerations, no matter what scheme for biomass growth is envisaged, and these in turn depend greatly on geographical location and climatic limitations. Inputs of solar energy, water, CO_2, nitrogen, phosphorus, and various minerals all contribute to plant growth and these, allied to conditions of temperature, O_2 concentration, soil quality, possible weather hazards, pollution, etc. all have a direct bearing on overall plant productivity.

With respect to global primary productivity, computer-generated models have estimated between 100×10^9 t and 125×10^9 t dry matter/year terrestrial productivity, plus between 44×10^9 t and 55×10^9 t dry matter/year produced in the world's oceans [40]. The geographical variation is naturally large, and Table 3.1 presents percentage productivities according to land type. The estimated mean efficiencies of terrestrial and marine photosynthesis are 0.3 per cent and 0.07 per cent respectively, while the overall virtual maximally attainable efficiency of 5.3 per cent for land plants was shown in Section 1.2. As an illustration of what can be achieved given adequate insolation, temperature, and

Table 3.1 Breakdown of world's primary productivity [3].

	Net productivity (% of total)
Forests and woodlands	44.3
Grassland	9.7
Cultivated land	5.9
Desert and semi-desert	1.5
Freshwater	3.2
Oceans	35.4

water supply by various crops in different parts of the world, Table 3.2 displays short-term yields and photosynthetic efficiencies. C_4 species are identified and these, growing in sub-tropical environments, may produce in excess of 400 kg/ha day, to provide the highest growth rates to be encountered under optimum conditions. In temperate regions like much of western Europe a summer maximum of 200 kg/ha day is possible.

Loomis and Gerakis [44] discovered a strong latitudinal dependence between productivity rates of C_3 and C_4 crops in agricultural systems, with the latter excelling at low latitudes but the former superior above 40–50°. There is a good correlation here with peak insolation levels and the occurrence of chilling temperatures to give a genotype-environment interaction at the crossover levels.

The high short-term productivities of Table 3.2 however are rarely, if ever, maintained over the growing season, even less so the complete year. In western Europe C_3 species like temperate forage grasses may produce 20 t/ha yr, at a mean photosynthetic efficiency of approximately 1 per cent, given sufficient water and soil nutrients. With adequate fertilizer input cassava yields of over 30 t/ha yr have been reported in the wet tropics, while nearly 40 t/ha yr of Malaysian oil palm has been achieved at a 1–1.5 per cent efficiency. C_4 species, with adequate water supplies, growing in tropical and sub-tropical regions provide the highest growth rates with sugar cane in Hawaii at 64 t/ha yr (2 per cent efficiency) and the tropical Napier grass at over 80 t/ha yr in Puerto Rico and El Salvador (2.4 per cent efficiency). All the above values are expressed in terms of dry matter, as are those presented in Table 3.3 below. The mean photosynthetic efficiency of many crops on an annual basis compared to their daily high growth rate periods may then be clearly observed.

The yields obtained in Tables 3.2 and 3.3 are those of planned agricultural operations in the main, whereby the desired product, be it food, fibre, etc. may be maximized for its production rather than whole plant dry

Table 3.2 High short-term productivities and photosynthetic efficiencies of selected plants [41–43].

Crop	*Location*	*Short-term high yield (g/m^2 day dry wt)*	*Photosynthetic efficiency*
Temperate:			
Tall fescue	UK	43	3.5
Ryegrass	UK	28	2.5
Ryegrass	Netherlands	20	1.9
Cocksfoot	UK	40	3.3
Potato	Netherlands	23	2.5
Sugar beet	UK	31	4.3
Sugar beet	Netherlands	21	2.5
Kale	UK	21	2.2
Barley	UK	23	1.8
Barley	Netherlands	18	1.7
Wheat	Netherlands	18	1.7
Peas	Netherlands	20	1.9
Red clover	New Zealand	23	1.9
Maize (C_4)	UK	24	3.4
Maize (C_4)	Netherlands	17	2.1
Maize (C_4)	New Zealand	29	2.8
Maize (C_4)	Iowa, USA	28	2.2
Maize (C_4)	Kentucky, USA	40	3.4
Sub-tropical:			
Alfalfa	California, USA	23	1.4
Potato	California, USA	37	2.3
Pine	Australia	41	2.7
Cotton	Georgia, USA	27	2.1
Rice	NSW, Australia	23	1.4
Rice	South Australia	23	1.4
Algae	California, USA	24	1.5
Sugar cane (C_4)	Texas, USA	31	2.8
Sorghum (C_4)	California, USA	51	3.0
Maize (C_4)	California, USA	52	2.9
Tropical:			
Cassava	Tanzania	17	1.7
Cassava	Malaysia	18	2.0
Palm oil	Malaysia (Complete yr)	11	1.4
Rice	Tanzania	17	1.7
Rice	Philippines	27	2.9
Bullrush millet (C_4)	Northern Territory, Australia	54	4.3
Napier grass (C_4)	El Salvador	39	4.2
Sugar cane (C_4)	Hawaii, USA	37	3.8
Maize (C_4)	Thailand	31	2.7

Table 3.3 Annual productivities and photosynthetic efficiencies of selected agricultural crops [41].

Crop	*Location*	*Yield (t/ha yr dry wt)*	*Photosynthetic efficiency*
Temperate:			
Ryegrass	UK	23	1.3
Kale	UK	21	1.1
Potato	UK	11	0.5
Potato	Netherlands	22	1.0
Sugar beet	UK	23	1.1
Sugar beet	Washington, USA	32	1.1
Wheat (spring)	UK	5 (grain)	0.2
Wheat	Washington, USA	12 (grain)	0.4
Wheat	Washington, USA	30 (grain)	0.1
Barley	UK	7 (grain)	0.3
Rice	Japan	7 (grain)	0.3
Sorghum (C_4)	Illinois, USA	16	0.6
Maize (C_4)	UK	17	0.9
Maize (C_4)	UK	5 (grain)	0.2
Maize (C_4)	Ottawa, Canada	19	0.7
Maize (C_4)	Japan	26	1.1
Maize (C_4)	Iowa, USA	16	0.5
Maize (C_4)	Kentucky, USA	22	0.8
Sub-tropical:			
Alfalfa	California, USA	33	1.0
Sugar beet	California, USA	42	1.2
Potato	California, USA	22	0.6
Wheat	Mexico	18	0.5
Wheat	California, USA	7 (grain)	0.2
Rice	NSW, Australia	14 (grain)	0.4
Rice	California, USA	22	0.6
Bermuda grass (C_4)	Georgia, USA	27	0.8
Sorghum	California, USA	47	1.2
Maize (C_4)	Egypt	29	0.6
Maize (C_4)	California, USA	26	0.8
Tropical:			
Oil Palm	Malaysia	40	1.4
Sugar beet	Hawaii (2 crops)	31	0.9
Cassava	Tanzania	31	0.8
Cassava	Malaysia	38	1.1
Rice	Northern Territory, Australia	11 (grain)	0.2
Rice	Peru	22	0.7
Rice & sorghum (C_4) multiple cropping	Philippines	23 (grain)	0.7
Sorghum (C_4)	Philippines	7 (grain)	0.2
Sugar cane (C_4)	Hawaii	64	1.8

Table 3.3 (continued)

Maize (C_4)	Thailand	16	0.5
Maize (C_4)	Peru	26	0.8
Napier grass (C_4)	El Salvador	85	2.4
Napier grass (C_4)	Puerto Rico	85	2.2

matter yields as such. Where crops are grown for their energy content then dry matter yields overall do become more significant, and it becomes necessary to increase the energy output : energy input ratio accordingly. These factors require a situation where man must have a reasonable amount of control over the system with respect to species grown, fertilizer, inputs, irrigation etc. The only means of achieving this is by the energy plantation concept.

It would still be possible to harvest some forms of natural vegetation as photosynthetically-derived energy sources, but probably not on an economical basis. The inevitable low yields would automatically require the sequestration of larger land areas than would be necessary for a preconceived energy farm. Additionally, if environmental conditions were conducive to good plant growth then such areas would probably have direct competition as sites for conventional agriculture, and might even be interspersed with land specifically designated for food and/or fibre production. Harvesting and transportation costs of the naturally grown biomass would inevitably rise as a result, to the obvious detriment of such schemes. Furthermore, wild species are less controllable biochemically, physiologically, and genetically than are deliberately developed strains, so yield improvement would be more difficult to obtain. The question of environmental and ecological imbalances must also be considered, including factors of soil erosion, desertification, deforestation, and the possible harmful consequences for the local natural flora and fauna. It would therefore appear that the utilization of natural vegetation as an energy source is severely limited, and must generally take second place to the energy crop plantation. Species selection can then be performed, with subsequent attempts to increase photosynthetic efficiency, reduce photorespiration, enhance soil biological nitrogen fixation to reduce fertilizer input, control pests, and possibly enrich the atmosphere with CO_2 [45].

An economic cost estimate has been performed of such a scheme, as yet hypothetical, by researchers at the Stanford Research Institute (SRI), California [46]. The biomass crop is envisaged as consisting of a conglomerate of annual and perennial species selected or developed for a high yield approaching 68 t/ha yr dry matter over an area of 4000 ha. Annual crops might include sunflower and kenaf for multiple-cropping, while sugar cane, sorghum, and forage grasses could constitute the perennial types, which could then be harvested more than once during the growing season. A subsequent energy budget has been postulated and will be discussed in Section 5.2 [47].

South-east California with south-west Arizona is cited as the primary potential region for the location of biomass plantations in the United States. Its particular advantages are plentiful good quality, publicly-owned land, the highest solar radiation intensity and most mean annual hours of sunshine in the entire USA, the longest growing season, very mild winters with high minimum monthly temperatures from November to February, the most rapid insolation increase in the spring (essential for starting off each year's first crop well), and, not least, a rapidly expanding economic base. Nevertheless, on the debit side is the extreme water deficiency of the area, with current annual consumption already exceeding the rate of replenishment at less than 200 mm annual rainfall. Interbasin water transfer from areas of water surplus would be a massive undertaking, the annual volume needed being 62×10^6 m^3, entailing a significant added energy input to the system.

The adjoining regions of eastern Arizona and New Mexico are also under consideration, but relatively more severe winter temperatures endanger the potential for year-round production, and the annual rainfall here would only provide just over 20 per cent of the required water supply.

Moving eastwards, southern-central Texas is thought to be the second favoured zone for fuel crop plantations. There are excellent winter temperatures, vast upgradable range lands, and deep, fertile soils, but drawbacks include a shortage of public land, comparatively low springtime solar radiation, only moderate annual solar radiation and mean annual sunshine compared with southern California, Arizona, and New Mexico, and the potential for catastrophic weather hazards. Forty per cent of the necessary water could accrue from rainfall, and there is a proposed scheme to import 155×10^{10} m^3 water per annum from the Mississippi River into the Texas water system in any event.

The south-eastern United States of Georgia and much of Florida are apparently water-sufficient and also possess favourable solar regimes. The potential for energy plantations in this area would therefore seem good, although much of the land may well be privately owned and there could be fierce competition from food and fibre producing crops. Economics would certainly have a great bearing here.

On a global basis it would appear that India and the Far East, together with tracts of land in Brazil, Argentina, and Uruguay, and possibly south-eastern USA and equatorial Africa, have sufficient indigenous supplies of both rainfall and insolation to avoid either reduced biomass yields or the provision of additional energy intensive inputs. This is not to suggest that energy plantations could not be viable propositions elsewhere, however. Indeed, a conducive climate such as that of much of Australia (excepting the desert regions) holds much promise, and five principal crops have already been selected as possible biological materials which could conceivably make a significant contribution to indigenous energy supplies: namely, cassava, kenaf, elephant grass, *Eucalyptus* spp., and sugar cane.

Undoubtedly the most ambitious and spectacular fuels-from-biomass programme which is actually operating has recently been instigated in Brazil, whose vast land resources enable extensive cultivation of high yielding cassava and sugar cane crops. Projects are also envisaged utilizing sweet potatoes and babacu nuts, with ethanol being the final product following fermentation and distillation in each case. The ethanol is intended for use as a motor fuel constituent up to a level of 20 per cent, thus making appreciable savings in petrol consumption, and hopefully aiding the country's balance of payments. In 1976 Brazil was paying more for its imported oil than any other developing nation in the world.

Three million m^3 per annum of anhydrous alcohol are estimated to be required to meet the demand, necessitating the construction of a further 193 distilleries by the Brazilian Government. The state of Bahia has a proposed 288 000 ha scheme for growing cassava, while a 400 000 ha tract of land in the remote territory of Amapa has been sequestered for the cultivation of sugar cane. The annual 2000 mm rainfall in these areas is more than adequate, and Brazil anticipates being self-sufficient in fuel alcohol by 1984 at the latest. However, although cane distilleries are good net energy producers and, by virtue of the combustion potential of the fibrous bagasse residue, can be autonomous, cassava residues contain a higher proportion of water so that it is thought that an external energy supply will be needed to run the processes, even when net energy producers. Brazilian scientists are apparently divided on the value of cassava as an energy substrate, and this subject will be returned to in Chapter 5, when energy analyses and their implications will be discussed.

It is the United States, with its inherently vast land reserves and variety of climatic conditions allied to extensive research and development funds and expertise, that has probably the most available options open to it regarding biological energy production. One particularly interesting recent study compiled by the Mitre Corporation, Metrek Division and the Georgia-Pacific Corporation, Portland, Oregon (for the former US Energy Research and Development Administration (ERDA)) focused on the potential of intensive short rotation silviculture plantations as a feasible method for producing energy products. This comprehensive appraisal occupies fully six volumes covering biomass potential, land suitability and availability, cost analyses, conversion processes and costs, and the potential of forestry wastes [48–53].

The concept would involve the intensive management of a densely planted energy crop under short rotations. Selected coppicing species such as *Populus* spp, hybrid poplars, *Eucalyptus* spp, *Alnus* spp, sycamore and tulip poplar could give rise to dry weight yields of 12.5–32.5 t/ha yr with current technology. Much depends on the species and on the site, and it is estimated that these yields could be doubled in 25 years, but perhaps necessitating further inputs.

The ratio of energy captured to that consumed was thought to be in the range from 10 : 1 to 15 : 1 [54]. Combustion of the wood to produce electricity was considered to be economically more attractive than its conversion to methanol, ammonia, and certainly to pyrolytic oil. However, it was stressed that before such a scheme could be developed commercially, both biomass yield optimization and conversion technology improvements required further advances. The availability of suitable land is also a crucial factor, and although trees may be cultivated on marginal land rather than agricultural, such land is competed for by more conventional forestry products, lumber as well as pulp.

Besides terrestrial energy farming, the possibilities of both freshwater and ocean farming have received considerable attention in the USA. The growth of microscopic and macroscopic algae as well as the water hyacinth, *Eichhornia crasspipes*, are being investigated with generally improved yields over land-based biomass production. Leaders in the field of solar energy fixation by algal-bacterial systems are Professor William Oswald and his associates at the University of California, Berkeley [55–60]. Such schemes involve the use of oxidation ponds in which microscopic algae provide oxygen to enable the symbiotically associated aerobic bacteria to degrade the large organic molecules present in the incoming sewage to smaller molecules like CO_2 and NH_3, which are then utilized, in turn, by the algae (Figure 3.1). The bacteria occur naturally as a small percentage of the total organisms present, and so do not have to be introduced artificially. Sufficient solar radiation is available all year round for the operation to be virtually a continuous one, although yields are inevitably reduced in the winter months. A 10^6 litre plant at Richmond, California gives rise to annual algal dry weight yields of approximately 50 t/ha, with the predominating genera being the green *Chlorella, Scenedesmus* and *Euglena*. Mixing is required to prevent cellular sedimentation, but the limiting economic factor is the harvesting procedure. At a pond concentration of 200 mg/litre it is estimated that the minimum power requirement per t dry wt algae is 2.7 Mwh(e) for the centrifugation stage. The addition of a flocculating

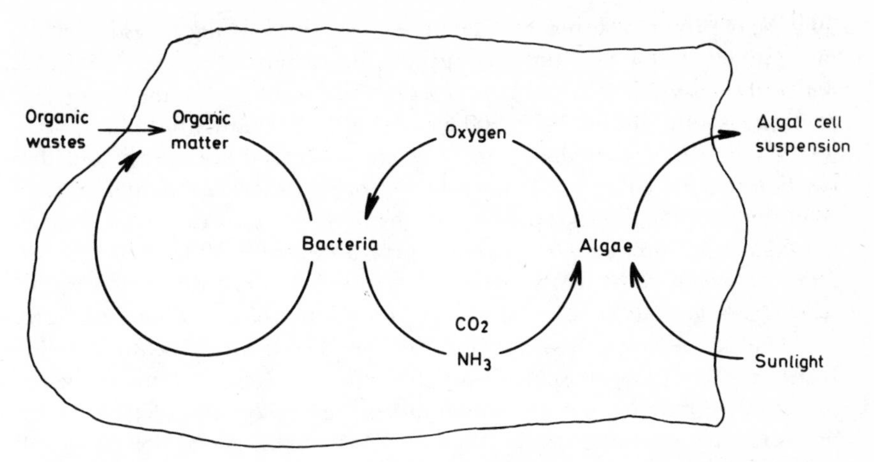

Fig. 3.1 Algal-bacterial growth cycle on organic wastes (after Oswald [59]).

agent such as alum would necessitate considerably less energy expenditure, although the algal product would thus be contaminated to the extent of increasing the cellular non-volatile matter percentage, perhaps to 14 per cent. Alternatively, the combination of growing filamentous genera such as the blue-green *Spirulina* sp. or *Oscillatoria* sp., with physical separation through screens would greatly reduce harvesting costs. Both algae are readily fermentable, though *Spirulina maxima* has shown little inclination to grow on raw sewage [60], but has been propagated in oxidized wastes [61]. It is thought that *Oscillatoria* would prove more favourable in an integrated system involving waste treatment because of its lower requirements for salt and alkalinity. However, it remains to be seen whether species control is feasible, and, at the time of writing, the photosynthetic efficiencies and productivities of these algae are still unknown. *Oscillatoria* and the colonial alga, *Microactinium* are apparently amenable to removal by passage through rotating microstrainers at reduced energy expenditure, but these genera are placed only sixth and seventh in abundance within oxidation ponds [62]. This underlines the advantages to be gained by species control, but the dynamics of algal populations are complex. Environmental changes, in temperature for example, greatly and rapidly affect the succession of species which predominate at any particular time, but although such parameters as temperature and insolation cannot be appreciably controlled, size-recycling, aimed at increasing the population of organisms easily harvestable, together with nutrient limitations, show promise of being at least partially effective [62–64].

The mass outdoor cultivation of macroscopic marine algae for the purpose of providing a potential energy source is also being investigated in the United States, this time mainly on the eastern seaboard. Two red seaweeds, *Gracilaria* sp., and *Hypnea musciformis*, were grown and harvested continuously in tanks of up to 600 litre volume. The initial nutrient source was a mixture of secondary-treated waste water and sea water, and the location was Fort Pierce, Florida which provided excellent annual growth conditions, with the exception that the *Hypnea* cultures could not be maintained in summer when tank temperatures rose to 30°C. Both species attained yields consistently between 12 g and 17 g/m^2 day dry weight, which averages out at some 60 t/ha yr, dry matter [65]. It is interesting to note that the investigators, members of the Woods Hole Oceanographic Institute, Mass., in subsequent articles advocate such systems as being better suited for energy conservation rather than energy production. Thus, instead of undergoing anaerobic digestion for methane generation, the algal biomass would be better deployed as protein producers, providers of fertilizers, drugs, and colloids, and as integral components of advanced waste-water treatment and waste recycling aquaculture processes [66–69]. The rationale here is that the resulting reduction or even replacement of more energy intensive conventional processes will give rise to quite significant, if not substantial, energy savings. Such a paradigm extends the concept of the energy plantation further to that of the energy conserving plantation which, in many situations, may be more appropriate, and indeed more successful!

Returning to the west coast, but continuing with an algal theme, the by now well-known 'ocean farm project' carried out off the Californian coast cannot be ignored, depite problems of seemingly fluctuating financial backing and sponsorship difficulties. The objective is certainly not lacking in ambition, however, namely to utilize the readily available solar energy and currently nonproductive ocean surface and depth nutrients for the production of the giant kelp, *Macrocystis pyrifera*. This, in turn, is primarily converted to methane gas, with the ensuing by-products used as fertilizer, feed supplements, and various industrial products like potassium and sodium salts, sugar residues, and gum materials.

The kelp is envisaged as growing on a system of polypropylene submerged supporting lines and buoyancy-control structures over an area of above 8000 hectares and at an ocean depth of 12–25 m. CO_2 and water would be acquired from the oceanic surface layers, with other nutrient requirements being provided by a system of recycling process waste and bringing up cool, nutrient-rich water from depths of 300–700 m. Photosynthetic efficiencies are estimated to be in the region of 2 per cent, with annual dry organic matter productivity ranging from 50–80 t/ha, excluding ash [70,71], but no figures have yet proved its net energy potential.

Finally, in this section devoted to energy farming, mention must be made of the once notorious water hyacinth, *Eichhornia crasspipes*. At present, many of the world's major rivers between latitudes 32°N and 32°S are infested and occasionally clogged by this highly prolific plant, which has the capacity to double in numbers, via vegetative reproduction, within 8–10 days in warm nutrient-enriched waters. The massively solid floating mats which form can produce up to 212 t/ha yr on a dry weight basis owing in part to their high levels of nutrient uptake. In an analogous manner to the Californian algal-bacterial systems for sewage treatment, water hyacinths can perform the same function, with the simultaneous production of huge quantities of biomass with great potential for conversion to substitute natural gas (SNG), good organic fertilizer, and quality animal feed supplement. Thus, the purposeful cultivation of this plant within domestic sewage lagoons and canals can be of great benefit in many ways. A hectare of *Eichhornia crasspipes* grown on sewage nutrients may yield 0.9–1.8 t dry matter each day, and, although optimum growth occurs in tropical and subtropical regions, there are distinct possibilities that its zone of habitation could be increased by use of raw sewage heat, transparent canopies, or thermal waste from industry and power plants [72].

An economic evaluation for the production of 5.7 million standard m^3/day of methane from water hyacinth cultivation has been performed by the Pratt and Whitney Aircraft Company in the USA, with quite favourable results [73]. The postulated growth area for this particular operation consists of 344 units, each incorporating two serpentine canals nearly 40 km long, giving a total canal length of some 26 400 km. Each hyacinth growth unit contains a collection/feed station with a chopper and pumps to transport a 10 per cent wet hyacinth slurry in water via a pipeline system to the central processing station. Much of the water and some of the chopped hyacinth is recirculated back to begin the propagation cycle once more. A total cycle time of two months is proposed, and the water used for transportation of the plants, together with excess water from the digesters, returns to the canals along the pipeline system. The three conclusions of the study are (1) that the total process is an attractive one due to the hyacinth's high propagation rate and low economic value, (2) the process is renewable and (3) the economic viability is assured even over the short term.

In the main, all the above systems are either at the conceptual or pilot-plant stage, with the exception of the Brazilian alcohol venture. Further improvements are thus likely as these various projects develop and there is considerable room for manoeuvre. It would seem that an integrated approach to the energy plantation concept would be both economically attractive and desirable, whereby food, fibre, fertilizer etc., are produced where possible as valuable by-products in addition to the primary energy output. However, a certain amount of waste is almost inevitable where biological systems are concerned, and the utilization of some of these as potential energy providers is discussed in the following section, although the net energy of many of these systems has yet to be subjected to critical analysis.

3.2 Biomass wastes

Conservation is, or should be, a priority, and it follows from this that the elimination of waste products, or products classified as waste, is also a priority. Perhaps, when it comes to considering biological processes, 'waste' is becoming an obsolete term in any event, and 'by-product' is a more accurate description, since virtually all organic matter potentially can be utilized in some way. Often the reason it is not is purely one of economics. Table 3.4 presents a breakdown of the annual quantities of organic wastes in the United States.

In the USA, cellulose, primarily in the form of paper, now accounts for more than 50 per cent of municipal solid waste each year [75], while in the UK domestic refuse amounts to some 18×10^6 t per annum, of which 58 per cent is organic (38 per cent paper and cardboard and 20 per cent vegetable and foodstuffs) [76]. About 6.5×10^6 t of waste paper, packaging and boards are discarded, yet less than 0.35×10^6 t are collected from domestic sources for repulping. Another 1.75×10^6 t are salvaged from trade premises, printers and paper mills.

Of all by-product materials produced in agriculture, straw probably has the greatest potential on a global scale as an energy source. Indeed, it has

Table 3.4 Source and quantity of solid organic wastes in the United States [74].

Source	*10^6 t dry wt/yr*	*10^6 t wet wt/yr*
Agricultural		2129*
Cattle	198	
Poultry	8	
Hogs	10	
Forestry	181	
Urban		232
Domestic	58	
Municipal	20	
Commercial	38	
Industrial		100

*Agricultural wastes include large volumes of vegetation wastes which are not collected and not included in the subheading. Forestry wastes are approximately equally divided between those accumulated at the mill and those scattered throughout the cutting areas.

been estimated that improved collection methods could realize an annual worldwide straw potential exceeding a billion tonnes, of which 10 per cent could be collected in the USA, thus making straw easily the most abundant of the non-woody plant materials [77]. Furthermore, the European Economic Community's 'Energy from Biomass' programme is concentrating on the potential use of straw as a fuel, initially in Denmark, France, and West Germany, where the combined annual total produced approaches 60×10^6 t (dry wt). Within the UK much of the cereal straw produced has merely been burnt in recent years, resulting in a loss of 3.4×10^6 t out of a total of 9×10^6 t (dry wt) each year [78, 79]. However, in 1976 much smaller quantities were squandered due to the then favourable market for straw utilization as an animal feedstuff. This, in turn, was a reflection of increases in the prices of other feedstuffs, and serves to illustrate the point made earlier that it is largely economics that makes a secondary product a waste, and vice versa.

In the United States, wood waste is frequently available at a rate of from 50 to several hundred tonnes a day [14]. Wood may contain 60 per cent fermentable sugars, and have a lignin residue of 20–30 per cent which can be used as a combustible by-product in the production process for ethanol manufacture (see Chapter 4). Softwoods (conifers) generally have a higher lignin content than do hardwoods.

World production of sugar cane bagasse in 1974 was around 44×10^6 t or more, used newsprint and other waste paper amounted to 133.5×10^6 t in 1972 [80], but the only cellulosic materials used commercially at the time of writing for significant industrial alcohol production are sulphite liquors [10, 81]. However, increasing use of the sulphate pulping, or Kraft, process will considerably reduce the amount of sulphite liquor available, so much so that the principal North American yeast production plants based on this as a substrate have been forced to close [82].

Probably the most comprehensive feasibility study of agricultural waste utilization as a prospective energy source is that carried out jointly by the Mitre Corporation and Georgia-Pacific Corporation in the USA, and mentioned in the previous section. This particular study is in addition to that outlined on the silvicultural farm concept, and concentrates on the by-products obtained from the growing and harvesting of commercial timber and its conversion to conventional forest products, namely the forest and mill residues. In general, the residues generated in the manufacture of forest products are used as an energy source within the industries themselves, which are about 40 per cent energy self-sufficient as a result. However, the residues, especially logging residues, formed via the cultivation and harvesting of timber are rarely utilized for energy production due to the high expense of collection, transportation, and handling, although this situation may well change where energy prices rise in real terms. The prices of wood residues fluctuate greatly both with time and location, but the potential for their conversion to SNG, for generation of electricity, and for production of process steam are certainly there [48, 53].

Many non-woody fibrous raw materials are used in the papermaking industry, but only a small percentage of the estimated global production listed in Table 3.5 is actually used for this purpose, and thus there is a large potential for energy production from these sources. An important consideration, as ever, is the spatial distribution of such crop fibres, for generally the higher this density then the less energy and money which need to be expended on their collection and transportation. Some data on annual collectable yields per hectare are provided in Table 3.5, as well as overall available quantities.

Much of the current wastage of these substances, particularly in the case of straw, could be overcome if an incentive policy for utilization were adopted. An integrated approach has been suggested so as to improve on the economics of operations carried out in isolation from each other. In this way a whole series of products, including fuels, could be manufactured in a single industrial complex. Straw of the finest quality could initially be used for paper pulp, and hydrolysis of second-grade material would then allow fermentation of the resulting sugars to ethanol, and at the same time give rise to high-protein yeast and cattle feed. Ethylene, furfural and other valuable chemicals might also be produced economically, and the final residue utilized as a fuel to generate power for running the plant [83]. This kind of scenario appears infinitely more worthwhile than a continuation of straw burning to dispose of the surplus, since this also leads to an obvious pollution of the environment. Additionally, the energy content of straw, 16 200 MJ/t, is appreciably greater than the energy cost of baling, carting, and transporting the straw, provided that the distance is not unreasonably long. A figure of 540 MJ/t is quoted for one Australian location [84]. Nevertheless, within a non-integrated strategy it would be difficult to imagine straw competing as a fuel precursor in the immediate future, except in certain on-site operations more particularly in the vast rural areas of the Third World.

Livestock, and indeed human wastes, also have much potential as an energy source in developing countries especially. Via anaerobic digestion, the wastes are treated, with the simultaneous evolution of methane fuel and the retention of nitrogen and other vital plant nutrients in the remaining sludge, which can then be recycled as a fertilizer. The prevalent practice of burning firewood and farm wastes in Third World regions to meet energy and fertilizer demands is not an attractive one. This can rapidly result in problems of deforestation on the one hand and loss of soil nutrients on the

Table 3.5 Estimated availability of specific non-wood fibrous raw materials, 1972, and estimated annual collectable yields per hectare [77].

Raw material	*Global availability ($\times 10^6$ t dry wt)*	*Collectable yields (t dry wt/ha)*
Sugar cane bagasse	55	5.0–12.4
Wheat straw	550	2.2–3.0
Rice straw	180	1.4–2.0
Oat straw	50	1.4–1.5
Barley straw	40	1.4–1.5
Rye straw	60	2.5–3.5
Flaxseed straw	2	0.5–1.0
Grass seed straw	3	
Subtotal, straw	*885*	
Jute	4.425	
Kenaf and roselle	1.674	1.5–6.2
Subtotal bast fibres	*6.099*	
Sisal	0.648	
Abaca	0.092	0.7–1.5
Henequen	0.164	
Subtotal leaf fibres	*0.904*	
Reeds	30	5.0–9.9
Bamboo	30	1.5–2.0 (natural) 2.5–5.0 (cultivated)
Papyrus	5	20.0–24.7
Esparto grass	0.5	
Sabai grass	0.2	
Cotton staple fibre	13.5	0.3–0.9
Second-cut cotton linters	1.0	0.02–0.07
Estimated total	*1027.203*	

other. 'Biogas' generation is a versatile process ranging in scale from the village-type 'Gobar' system of rural India and the Far East for only a few animals, to a much more intensive farming complex where relatively large volumes of organic matter are continually available. The former depends greatly on a hot climate and manual labour for its functioning, while the latter, although it would be energetically more efficient with higher ambient temperatures, is suitable for quite temperate countries [85], and substitutes fossil fuel inputs for a large manpower requirement. The process itself and the energy balances involved will be discussed in Chapters 4 and 5 respectively.

The amount of manure which is available for bioconversion is prodigious, but again collection and transportation greatly affect the energetics and economics of any proposed operation. Nevertheless, in order to avoid contravening the increasingly stringent anti-pollution laws of many Western societies this waste has to be treated in any event, and not merely dumped as might have occurred in the past. Thus the whole process may be construed as being one of primarily waste-treatment, with the additional production of good quality fertilizer, and the methane itself considered as something of a bonus [86, 87]. In the USA the total annual livestock waste amounts to some 216×10^6 t dry wt, and in the UK approximately 46×10^6 t [85]. If the estimated human solid excreta were also taken into account, the UK total would rise to around 52×10^6 t, and it has been suggested (perhaps optimistically) that the methane obtainable from this via anaerobic digestion could replace up to 25 per cent of the nation's annual gas consumption [88]. By analogy one may assume that the US figure including human faeces would increase to about 240×10^6 t.

Any methane produced from livestock wastes in the developed world will almost certainly be utilized on-site, as it were, in agricultural practice, at least in the foreseeable future. However, in the Third World, and especially in India, China and other Far East countries such organic wastes can go a long way to providing much of the energy required by a large percentage of the population. In India, for example, 75 000 biogas plants are already in existence and there are 235 million cattle in the country (compared to 13.5 million in the UK). For a typical village of population 500 the number of cattle would probably be around 250 and the amount of dry dung collected from them approximately 180 t/yr. This would be sufficient to meet the fuel requirements of the village population, on conversion to methane, except for two or three winter months, when there will be a shortfall of some 20–40 per cent. However, with further optimization of the anaerobic digestion process this shortage could conceivably be reduced to zero. As previously stated such a scenario is infinitely preferable to burning timber and farm wastes for an energy supply, and emphasizes that there need be no such term as 'waste' given the appropriate innovative technology, and, equally important, the willingness to use it [89, 90].

The satisfactory disposal of urban refuse is becoming increasingly difficult owing to its potential harmful effects on the environment. Figures for the amounts of such materials produced annually in both the UK and the USA were briefly mentioned at the beginning of this section. Suffice to say that the problem is a large one now, and will undoubtedly be amplified during periods of continued economic growth. However, much of this categorized waste can be recycled, particularly metals and paper, while there is also a utilizable protein content for animal feed, as well as the possibility of energy production. Which schemes, if any, are adopted in a particular situation depend largely, and inevitably, on the economics involved. For instance, incineration of refuse achieves its prime goal of volume reduction by 90 per cent, but the process can be extended, at extra cost, to provide energy in the form of electricity for example. At present the economics of such recuperative incineration are not particularly attractive, however, and neither are the alternative energy production routes via anaerobic digestion or pyrolysis, which requires heating the refuse in the absence of oxygen to temperatures above 500°C [76]. Nevertheless, various projects aimed at energy recovery from municipal solid wastes are underway in the United States which add up to over 1000 MW generating capacity, derived from 44 000 t of refuse each day [91]. If it were possible to convert all UK domestic refuse to methane through anaerobic digestion then this would satisfy around 5 per cent of the nation's gas demand, but at present 96 per cent of this waste is sanitary landfilled and so there is a long way to go before a significant impact could be made from this 'energy from waste' concept, especially when the costs are also taken into account [92]. The various processes and economic considerations will again be discussed in greater detail within Chapters 4 and 6 to follow.

Chapter 4

The bioenergy conversion technologies

4.1 An overview

This section is essentially descriptive, and covers the conversion routes whereby organic matter energy is transformed into a product usable by man as a source of energy. The various processes discussed are generally either being operated at present on a large scale or are at the pilot plant stage. What might be termed the more futuristic systems, such as direct hydrogen production by plants are covered later in Chapter 7. Not all the conversions are brought about by wholly biological means, but in all cases the substrate is biomass of some form or other, such as wood, other plant material, animal waste, urban organic refuse, etc. Figure 4.1 diagrammatically presents the principal routes and products involved, the processes themselves being categorized into three fundamental types, namely combustion, production of fuel(s) from essentially dry biomass by chemical means, and aqueous processing. Which method is selected in a given situation firstly depends on the product desired. Since a particular fuel may be obtainable via several different means there is some scope for manoeuvre, and thus the secondary criteria may then be brought into the decision-making process. Chief among these is the biomass' water content which, if high, may preclude all but aqueous processing, owing to the energy requirement for drying the material artificially often being excessively large compared to the energy content of the product eventually formed. Natural, field drying cannot be considered in many parts of the globe because of

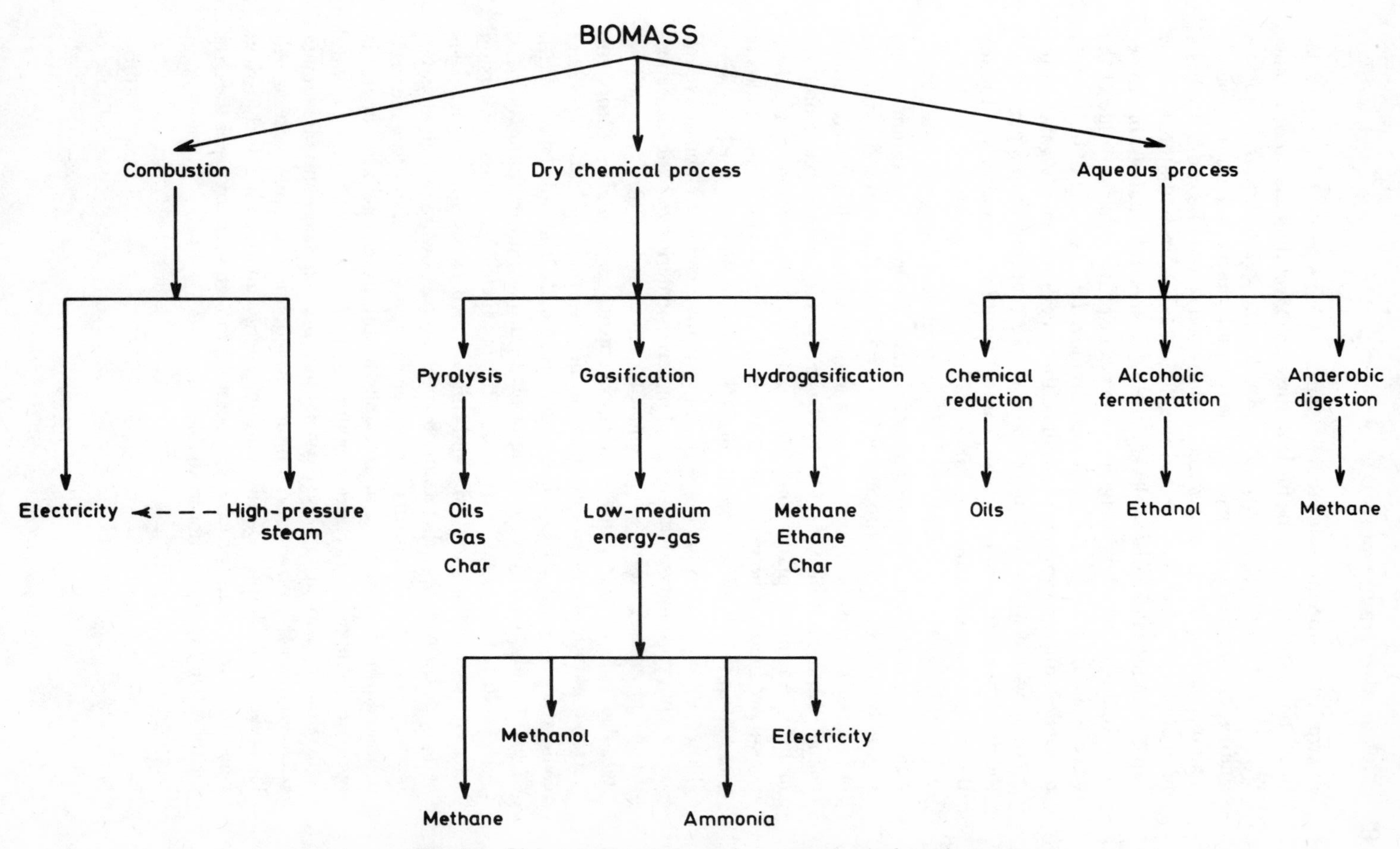

Fig. 4.1 Biomass energy conversion processes and products.

intermittent, unpredictable rainfall occurring during periods of good insolation, and would probably be futile in the case of wet biomass (above 70 per cent moisture content) in any case. Most crops, algae, and other biomass sources are in this category, but even freshly harvested timber usually contains around 50 per cent water. If this wood were to be combusted as a prerequisite to electricity generation, for example, then the heat energy input to reduce the moisture level to below the maximum allowable 15 per cent would be approximately 20–25 per cent of the calorific value of the dried wood itself. For each kg of water removed 3.5–4.5 MJ of heat would be needed, and so significantly reduce the overall net energy gain. At the extreme, with a water content of above 80 per cent, less than 5 per cent of the dried biomass' original energy content would remain after subtracting the energy consumed in evaporation [27].

Other important considerations are those of fluctuating substrate availability and the associated problems of storage. In order to maintain a high overall load factor at the processing plant it is desirable that a suitable mix of biomass sources may be obtained and dealt with as climatic conditions change throughout the year. On this basis it would be desirable to produce multispecies energy plantations, for instance, with approximately identical overall productivity no matter which season is in progress. A huge influx of material in the summer months might necessitate the need for large-scale storage by embarrassing the conversion plant's capacity; and it is well known that most harvested biological materials deteriorate badly on exposure to the vagaries of the environment for any length of time, whether by auto-decay, by the activities of living organisms (particularly saprophytic bacteria and fungi), or simply through the absorption of moisture. There could thus be need for storage under reasonably controlled conditions, though ideally, with correct sizing of plant, this should be minimal.

It would not be so difficult to store the products of biomass conversion (electricity, of course excepted); and similarly, where transportation is concerned, because of the bulky nature of most biomass owing to high moisture levels, if a choice has to be made between conveying either the raw material or the end-product fuel any appreciable distance, then the latter course is more desirable on energy grounds alone.

Economics also inevitably play a large part in decisions of biomass selected, process utilized, and product formed, and these implications will be covered in Chapter 6. In addition, energy analyses are included in Chapter 5 for as many processes as possible, though in many cases the data for these are as yet just not available.

4.2 Anaerobic digestion

Since this book is primarily concerned with energy production via biological systems it would seem appropriate to begin the various conversion process descriptions with the aqueous operations shown in Figure 4.1, since these are almost entirely biologically-mediated transformations.

4.2.1 Methane from manure

Anaerobic digestion is such a process, and has a wide range of applications. The arrangement of a digester plant is in its most simplified form for the treatment of livestock waste. Thus, the manure from, say 150 cattle may initially be macerated in a 4 m^3 volume tank, from whence the slurry, containing about 9 per cent solids on a dry weight basis is passed to a similarly sized mixing vessel, prior to being pumped into the digester tank itself. Often, two digesters are operated simultaneously, and two 5 m^3 working volumes with a retention time of 15 days have been used in the UK, at a constant temperature of 35° C and with mechanical agitation applied to maintain homogeneity as far as possible throughout the vessel [93]. Savings in heating requirements may be effected by keeping the input tanks below ground, lagging both pipes and vessels, operating a

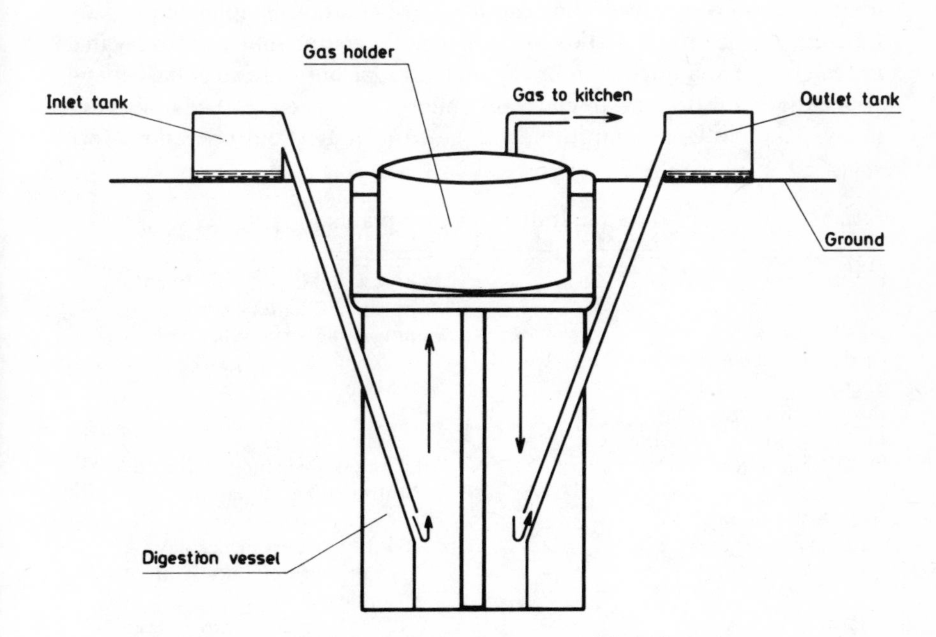

Fig. 4.2 Typical Indian subterranean 'biogas' generator.

reasonably efficient heat exchange-boiler system, and attempting to recover some of the output heat via a fitted heat exchanger.

The products of digestion consist of two streams. The 'biogas' stream contains methane and carbon dioxide in an approximate 2:1 volume ratio, the latter of which may be removed by various methods, including contact with monoethanolamine, in order to upgrade the fuel quality of the gas. Around 120 m^3 of 'biogas' with an overall energy content of from 2.4–3.1 GJ have been obtained during the above 15 day operation, but with improved conditions this quantity could be significantly increased. It must be stressed that in many ways the energy value of the evolved gas can be considered as a bonus, since the remaining sludge following digestion retains its nitrogen content to a value of 1.5–2 per cent as opposed to that of composted manure's 0.75–1 per cent N after losses [90]. Thus, in a way, the digestion process enhances the dung's fertilizer quality. Also, in the context of the Indian situation discussed in Section 3.2, the gas production is vital as a fuel since carbon present in dried manure, burns with only an 11 per cent efficiency in the traditional open fire, as opposed to the 60 per cent and above efficiency from methane obtained via 'biogas' plants [90].

Figure 4.2 represents a simple Indian 'biogas' plant, consisting of the digester, in which the fermentation takes place, and the gas holder, in which the evolved gas collects. The volume of both can be varied depending upon quantities involved and length of fermentation. Capital inputs are minimized, with the digestion vessel generally comprising a pit excavated in the ground and lined with brick, while the gas holder is steel-based, and floats above a water seal to maintain uniform gas pressure. The units have many possibilities for improvements both in design and operation since

Table 4.1 Relevant data for Indian 'biogas' plants [90].

Input of manure (dry wt)	approx. 2.8 kg/day per buffalo approx. 2.0 kg/day per cow approx. 0.8 kg/day per calf
Production of gas	At 15%, 0.18 m^3 of gas/kg of manure
Energy value of gas	20 MJ/m^3
Burning efficiency	60%
Effective (useful) heat obtained	12 MJ/m^3 of gas
Production of fertilizer	0.72 kg of dry sludge with 2% nitrogen/kg of manure
Gas consumption	
for cooking	0.34 m^3 per person per day
for lighting	0.125 m^3 per hour per lamp of 100 candle power
for motive power	0.425 m^3 per horse power hour

thus far in their history simplicity has been the keyword, with virtually no energy inputs for heating, stirring, etc. other than those provided by manual labour. To prevent corrosion, the steel gas-holder is painted annually, but again maintenance requirements are minimal. Since no attempt is made to acquire the optimum fermentation temperature, gas production rates are not constant, that for December being less than 50 per cent of that of a typical summer month. The application of solar, rather than fossil fuel, energy for heating the digester contents is now becoming quite a popular, and hopefully profitable innovation. Pertinent performance data for Indian rural 'biogas' plants are shown in Table 4.1.

4.2.2 Methane from algae

The bioconversion of algae into methane within an integrated system which also includes sewage treatment and disposal has partly been described in Section 3.1, concentrating on the algal growth and separation procedures. Energy analysis data for this particular scenario, along with anaerobic digestion using livestock waste and urban refuse will be presented in Section 5.4.

A large-scale methane-from-algae facility has yet to be operated, but experiments in California using six different algal species and incorporating a digester detention time of 20 days, a loading of 4 kg algal volatile matter (i.e. discounting mineral ash content) per m^3 digester volume, and a working temperature of 37°C, has resulted in gas volume ratios of 71 per cent methane to about 28 per cent carbon dioxide [60]. However, problems have been experienced with respect to the resistance of a certain proportion of the algal cells to enzymatic degradation, the build-up of ammonia within the fermenters, and the dewatering properties of the residual sludge. These residues could be recycled to the algal growth ponds to increase production, but since this sludge contains virtually all the nitrogen and phosphorus from the fermented algal biomass then its economic credit as a fertilizer would be lost, and it is doubtful whether this could in fact be justified. In such a delicately balanced integration each facet must be optimized with respect to the other parameters for the benefit of the whole. Thus, the resulting methane itself could be burned directly to produce power so that the products of combustion and the waste heat could be recycled, the CO_2 for instance being returned to enhance the algal growth phase. Carbon dioxide could also advantageously be accrued from a fossil fuel power plant, a suitable industrial process, or even a geological source [62].

It would appear that it is non-viable to utilize conventional anaerobic digesters here, with their inherently large capital investments in steel and concrete and significant operational energy inputs for heating and mixing.

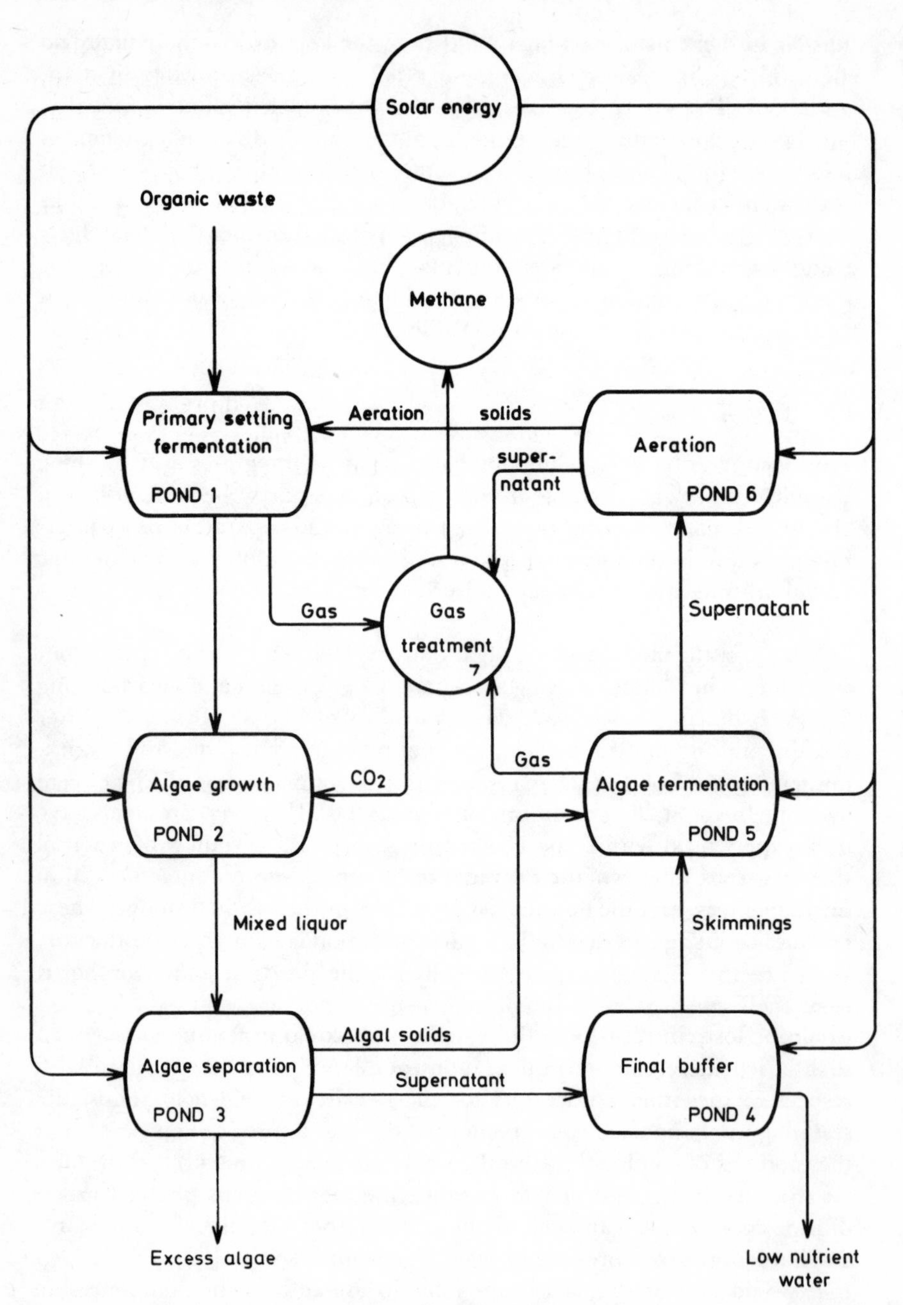

Fig. 4.3 Algal-bacterial solar energy fixation and conversion scheme (after Oswald [94,95]).

A solution may lie in the construction of anaerobic ponds fitted with floating plastic methane collectors, also serving as solar collectors, where the capital cost could be cut by over 95 per cent and heating and mixing reduced appreciably to enable allowable increased detention times to be operated. However, it must be stressed that the process *in toto* awaits further development before Professor Oswald's envisaged system shown in Figure 4.3 becomes a workable reality – although the time may not now be too far away [94, 95].

This scheme does include the recycling of the fermentation effluent, rich in ammonium nitrogen, phosphorus, carbonate, and trace minerals, to provide an optimum medium for the algae growing within Pond 2 of the figure. Following propagation, the algae should be concentrated prior to digestion at a solids level of 1–3 per cent, theoretically accomplished in Pond 3. The cells may then be fermented in the anaerobic Pond 5 to produce methane gas, CO_2, and ammonium so that the cycle can be repeated. The fermentation ponds themselves are envisaged to be around 9 m deep, heated and equipped to collect the total gas yield, which is finally scrubbed within block 7 of Figure 4.3.

4.2.3 The biochemistry and microbiology of methanogenesis

The entire above operation, an expansion of Figure 3.1 to include the anaerobic methanogenic component, is demonstrated schematically in Figure 4.4. Taking the right-hand side of this figure, and legitimately generalizing to include methane production from substrates other than algae, such as livestock waste, municipal solid waste, water hyacinths, etc., four basic processes may be distinguished as occurring within the anaerobic digesters. Primary hydrolysis involves the enzymatic conversion of insoluble organic compounds, such as cellulose by cellulase enzymes, to soluble organics. The second stage involves the fermentation of the stage one end-products (carbohydrates, proteins, lipids, alcohols, etc.) by non-methanogenic organisms to organic acids, predominantly acetic and propionic. These acids are converted by methanogenic bacteria into dissolved methane and carbon dioxide, which themselves finally undergo transfer from the liquid to the gaseous phase [87].

Several of the bacteria involved in the various stages of anaerobic digestion have yet to be adequately identified. Of those observed in the digestion of piggery waste, for instance, approximately 50 per cent have been found to be facultatively anaerobic streptococci, with *Bacteroides* spp. and *Clostridium butyricum* also present. *Escherichia coli* and other coliforms have been recognized in working domestic digesters, and *Methanobacterium* spp. along with *Methanobacillus* spp. are said to be the principal methanogenic organ-

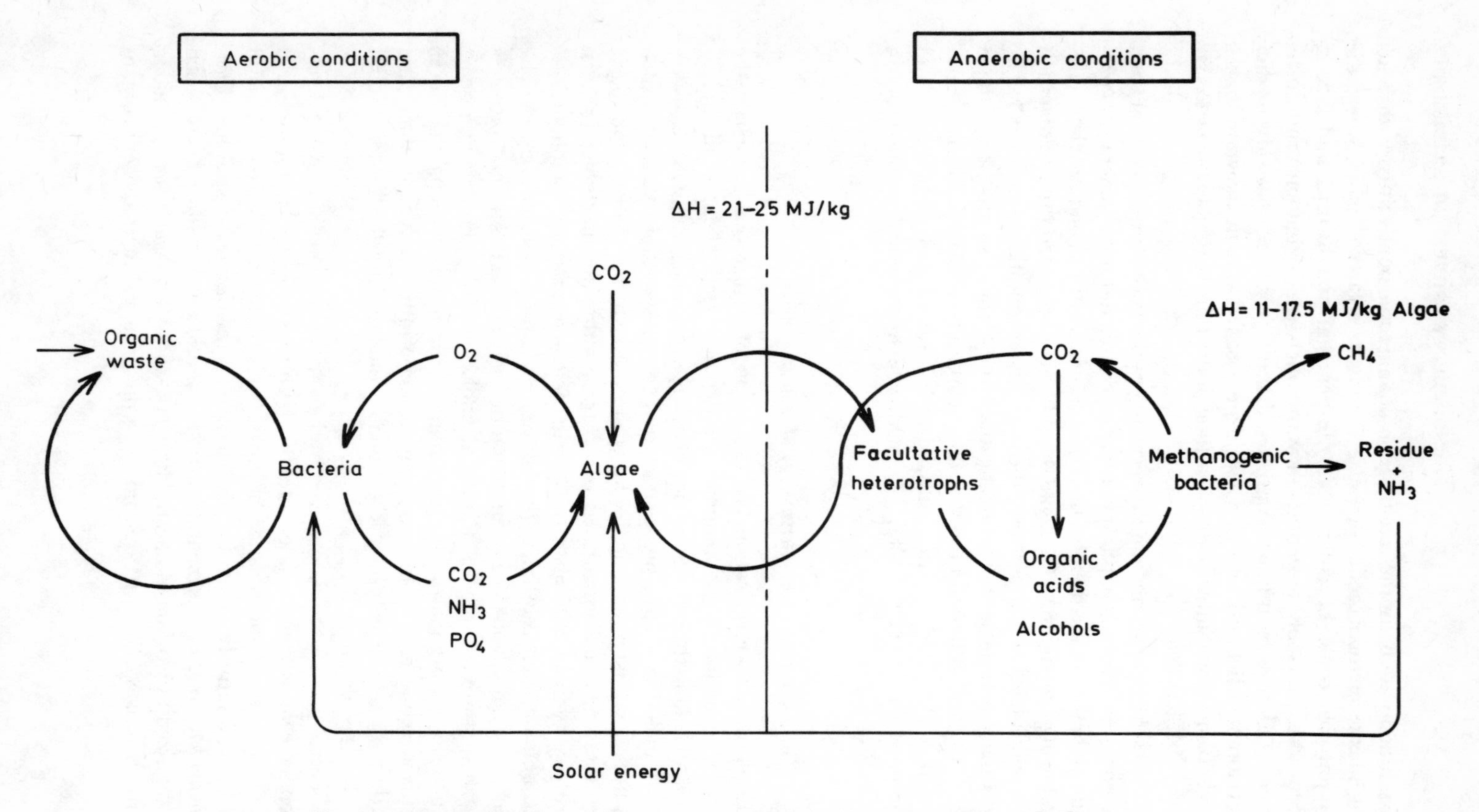

Fig. 4.4 Algal-bacterial growth cycle and methanogenesis (after Oswald [95]).

isms implicated, although the taxonomy of methanogenic bacteria has still to be satisfactorily organized [96].

The development of faster overall reaction rates would naturally be desirable, and the rate-limiting step must be identified and measures taken to increase the speed at which this stage operates. The methanogenic step is usually considered to be rate-limiting, but an experimentally tested hypothesis has tended to associate the rate-limiting stage more closely with the final transfer to the gas phase, and that additionally the metabolism of the methanogenic bacteria is actually inhibited by the product gases formed [97]. Faster, more efficient anaerobic digesters are envisaged to overcome this problem to a certain extent, and new designs are already being developed with this in mind (see Section 5.4), together with improving the overall energy balances involved.

In general, the overall reaction of converting organic material into carbon dioxide and methane can be represented as follows:

$$C_nH_aO_b + [(4n - a - 2b)/4]H_2O \rightarrow [(4n - a + 2b)/8]CO_2 + [(4n + a - 2b)/8]CH_4;$$

and where cellulose is the most common starting point, as in the case of the municipal solid waste discussed below, then the principal equation of interest becomes:

$$(C_6H_{10}O_5)_n + nH_2O \rightarrow 3nCO_2 + 3nCH_4.$$

However other complex compounds of wider differing chemical structures may also be found as a proportion of the microbial substrate. Aromatic polymers such as lignins and tannins can occur which are dismembered, often with some difficulty, into smaller aromatic components by the action of aerobic extracellular microbial enzymes. The ultimate benzenoid structures formed are eventually converted to methane and carbon dioxide by anaerobic methanogenic organisms including *Methanobacterium formicicum* and *Methanospirillum hungati* via a series of reactions such as the following:

1. $4\ C_6H_5COOH + 24\ H_2O \rightarrow 12\ CH_3COOH + 4\ HCOOH + 8\ H_2$
2. $12\ CH_3COOH \rightarrow 12\ CH_4 + 12\ CO_2$
3. $4\ HCOOH \rightarrow 4\ CO_2 + 4\ H_2$
4. $3\ CO_2 + 12\ H_2 \rightarrow 3\ CH_4 + 6\ H_2O$

This represents overall:

$$4\ C_6H_5COOH + 18\ H_2O \rightarrow 15\ CH_4 + 13\ CO_2;$$

demonstrating the biochemical complexity of the anaerobic digestion process, much of which has yet to be fully elucidated [98].

4.2.4 Methane from urban organic waste

The availability and amenability of urban solid waste for anaerobic digestion has been mentioned in Section 3.2 and will be referred to again in Section 5.4. A good example of what is potentially achievable in this area is given by the Dynatech R/D Company of Cambridge, Massachusetts and their bioconversion process to methane. A block diagram of the whole operation is presented as Figure 4.5 [99].

Initially the municipal waste delivered to the plant site is conveyed to the size reduction facility whose dual functions are to separate out the inorganic non-digestible material and to render the feed to the digesters more susceptible to solubilization. Separation may be effected either by a dry or a wet system, the former having the inherent advantage of allowing the desired water content of the digester feed to be selected. Both methods normally require primary shredding for particle size reduction down to 7.15–15 cm, ferrous metal is removed by magnetic separators, and fine grit and glass are separated out by passage through screens. The process stream is then fed into an air classifier in the case of the dry system, where the lighter, organic fraction is recovered, while the metals, glass, and heavier organic fraction may be treated further in a resource recovery system before being recycled. This differs from the wet procedure in which the waste is mixed with a large volume of water inside a hydropulper, where additional metal is removed, and fibrous cellulosic matter is obtained in a dilute suspension from the stream via a liquid cyclone. A penalty of dewatering is incurred when a concentrated cellulosic stream (>2–3 per cent) is preferred to be passed to the digesters.

Prior to digestion, nutrients, supplied by raw sewage sludge, and lime and ferrous salts for pH and hydrogen sulphide control, are added, along with liquid, conveniently available from the final liquid effluent, where needed following a dry separation. All these inputs are metered and blended in a mixing tank from which they are fed into a battery of digesters.

The digesters each consist of a large cylindrical tank with an associated stirring mechanism (turbine or propellor mixers, gas draft-tube mixers, or sludge recirculation pumps) to facilitate uniform digestion of the contents. Additionally, a floating cover is desirable for maintaining a constant pressure, as was the case for the digesters used in the agricultural practice outlined above.

The solids are firstly solubilized by bacterial enzymes, and the ensuing solubles undergo a reaction sequence as indicated for all such anaerobic digestion procedures. In the case of mesophilic operation, a temperature of 35–38°C is acquired by either injecting steam directly into the feed slurry streams or by preheating the aqueous stream in a heat exchanger with

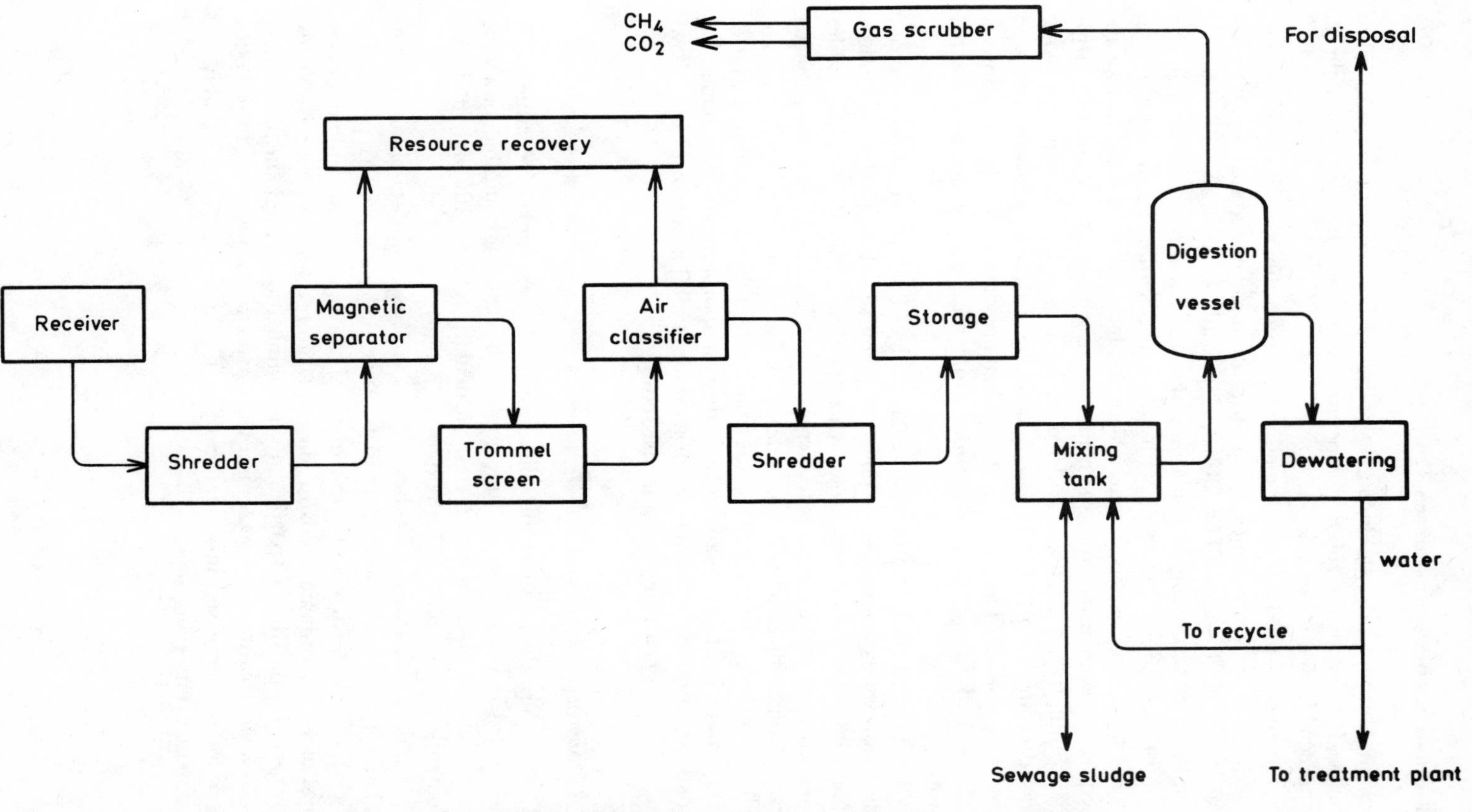

Fig. 4.5 Scheme for methanogenesis of municipal waste process (after Wise *et al.* [99]).

steam. During winter the incoming sewage sludge temperature may fall to below 10°C, with the solids' temperature even less, thus necessitating an appreciably greater heat input.

The overall products are a methane : carbon dioxide gas output of approximately 50 : 50 volumes, and a slurry stream of undigested organic matter ready for disposal. The CO_2 and possibly small traces of H_2S, are removed by scrubbing in a manner akin to that already utilized in the natural gas industry. Contact with monoethanolamine in an absorber can be used to perform this function, with the monoethanolamine being subsequently regenerated from the contaminating gases by heating in a stripper. Similarly the methane is dried by its accompanying moisture being absorbed in dry glycol, the glycol again being subsequently regenerated and recirculated. The ensuing energy balance data of Table 5.6. shows a conceived 4 TJ methane output/day from an input of 900 t municipal waste.

The effluent slurry contains water, non-digestible matter, undigested but digestible solids, solubilized organic compounds, and the bacterial biomass. This viable biomass is separated, where possible, from the non-digestibles and returned to the digesters to increase, or at least maintain, the previous rates of digestion and hence gas generation. Slurry disposal is finally accomplished by separating out the solids from the liquid and returning the latter to a sewage treatment plant to be dealt with appropriately. The remaining solids are ultimately incinerated or used for landfill as a moist cake following filtration, centrifugation, or just settling.

4.2.5 Summary

To sum up then, anaerobic digestion is an extremely versatile, well developed and comparatively uncomplicated technology suitable for a wide range of raw material substrates, particularly those with high moisture contents, as exemplified by the data in Table 4.2.

It should be stressed that these figures are based on data obtained from laboratory digesters operated at 30° C. Nevertheless, in modern plants, gas yields of 0.45–0.5 litre/kg organic material can be expected, and tank loadings of 4–5 kg organic solids/m^3 day in large reactors to maximize yields are certainly possible. In particular, the combination of methane production with waste treatment processes is a very attractive scenario, and this bodes well for the increased introduction of such schemes within a suitably enlightened conserving society.

Table 4.2 Gas yields from the anaerobic digestion of solid wastes [100].

	Total gas yield (m^3/kg dry solids)			*Methane yield (m^3/kg dry solids)*		
Solid waste	*Total solids basis*	*Volatile solids basis*	*% methane in gas*	*Total solids basis*	*Volatile solids basis*	*Half-digestion period (days)*
Municipal sewage sludge	0.43	0.60	78	0.34	0.47	8
Municipal sewage skimmings	0.57	0.63	70	0.40	0.44	–
Municipal garbage only	0.61	0.63	62	0.38	0.39	6
Waste paper only	0.23	0.26	63	0.14	0.16	8
Municipal refuse (combined, free of ash)	0.28	0.31	66	0.18	0.20	10
Abattoirs waste:						
cattle paunch contents	0.47	0.53	74	0.35	0.39	13
intestines	0.09	0.09	42	0.04	0.04	2
cattle blood	0.16	0.16	51	0.08	0.08	2
Dairy wastes, sludge	0.98	1.03	75	0.74	0.77	4
Yeast wastes, sludge	0.49	0.80	85	0.42	0.68	–
Paper wastes, sludge	0.25	–	60	0.15	–	–
Brewery wastes, (hops)	0.43	0.45	76	0.33	0.34	2
Stable manure (with straw)	0.29	0.35	75	0.22	0.26	19
Horse manure	0.40	0.44	76	0.30	0.33	16
Cattle manure	0.24	0.32	80	0.19	0.26	20
Pig manure	0.26	0.42	81	0.21	0.34	13
Wheaten straw	0.35	0.37	78	0.27	0.29	12
Potato tops	0.53	0.61	75	0.40	0.46	3
Maize tops	0.49	0.52	83	0.41	0.43	5
Beet leaves	0.46	0.50	85	0.39	0.43	2
Grass	0.50	0.56	84	0.42	0.47	4
Broom (25 mm cuttings)	0.44	0.45	76	0.33	0.34	7
Reed (25 mm cuttings)	0.29	0.32	79	0.23	0.25	18

4.3 Alcohol fermentation

The other principal aqueous conversion method is afforded by the familiar yeast fermentation route to alcohol production. Although butanol, acetone, and other low molecular weight organic compounds with a potential use as fuels may be obtained, the principal end-product normally derived for this purpose is ethanol. The envisaged shortfall in petroleum-based liquid fuel availability at a moderate price to the consumer could well result in a substantial increase in biological liquid fuel production before the turn of the century. There are numerous carbohydrate raw materials which may be ultimately converted to ethanol, although generally a pre-requisite to the alcoholic fermentation itself is the provision of a fermentable glucose substrate. Thus, varying degrees of pre-treatment are necessary, dependent upon the molecular complexity of the raw material starting point, and this aspect is patently reflected in the energy accounting procedures adopted subsequently in Section 5.5.

The ubiquitous nature of cellulose as the most abundant product of photosynthesis renders such substances as straw, wood, bagasse, and waste paper as obvious candidates for utilization as an ethanol source. However, these are comparatively 'dry' materials and might be exploited as energy sources to greater advantage via combustion, pyrolysis, or another of the non-aqueous treatments; but circumstances prevailing at any particular time might equally well dictate ethanol manufacture to be a top priority, as indeed occurred in Germany during the Second World War. In their case the Germans converted particulate wood material into ethanol by hydrolysis, using the dilute acid Scholler process, followed by conventional yeast fermentation and distillation. The physical inputs to, and energy analysis of, the whole operation are presented further on in Table 5.9. Generally, the process requirements using wood also apply to other cellulosic substances, with only minor modifications. A description of the German plant process follows, as an introduction to this sub-section, as being a fully operational concern, demonstrating the fact that, here at least, one is dealing with a proven, well known technology which actually provided a fuel both utilized by man and derived from photosynthesis and the ensuing biological, chemical and physical treatments required.

4.3.1 Ethanol from wood via acid hydrolysis

The Scholler Holzminden operation [14] involved six 50 m^3 digesters, constructed of steel and lined with acid-resistant tile. Digester diameters were each 2.4 m, and overall height about 13 m. At the top were steam and vent lines and a line for the injection of dilute acid, while the bottom was equipped with a filter cone and a quick-opening valve for discharging the resistant lignin residue.

To commence each run the digester was loaded with 9–10 t of sawdust, wood shavings and chips to a density of 180–200 kg dry wt/m^3. In order to attain this density the percolator was filled and subjected to a sudden steam shock to compress the charge, this being repeated until the desired loading was obtained. The wood was then heated with direct steam to 134°C, and a charge of dilute sulphuric acid injected at a temperature below this to be heated by steam from the bottom, so as to arrive at the required temperature. By applying steam to the top of the charge the solution formed was 'pressed' from the percolator, the whole procedure repeated approximately 19 times with 0.6 per cent acid temperatures increasing to a maximum of 184°C. The total amount of liquor acquired per percolation was around 120 t (120 m^3) from 10 t of woody raw material. The sugar concentrations formed were quite low, being in the region of about 3 per cent, and repolymerization of the monosaccharide entities could be kept minimal by maintaining the solution at 100°C for 12 hours, thus obviating the need for any post-hydrolytic treatment. Sugar yields have been experienced as high as 50 per cent by weight. A schematic diagram is presented as Figure 4.6. Figure 4.7 extends the hydrolysis stage, to include the subsequent fermentation and distillation stages, but at a somewhat simplified level. Figure 4.7 represents an improved operation of the Scholler process which has been more recently developed by the US Forest Products Laboratory to permit a semi-continuous procedure, reduce residence time, and thereby improve yields. In this Madison process wood biomass passes from a storage hopper to the digester, steam is introduced, the temperature rises to 125°C, and sulphuric acid is added along with a dilute prehydrolysate which has been recycled. Hot water is finally added and the temperature kept at 135–150°C for 30 minutes, the prehydrolysate is drained, both dilute sulphuric acid and steam at high pressure admitted for the main hydrolytic stage which continues for three hours, and at the end of which a temperature of 190°C is attained.

The solution is 'flashed' to a lower pressure, when any methanol and furfural present are separated in a distillation tower, the hydrolysate is neutralized with lime, and calcium sulphate precipitate removed for washing and sugar recovery. Ultimately the sugar is transferred to a yeast fermentation from which the dilute alcohol stream is concentrated to 95 per cent ethanol (190° proof) on passage to distillation columns, with pentose sugars remaining in the still bottoms.

Although the Scholler and improved Madison techniques are probably the most popular acid hydrolytic procedures, there are several others employing various reagents and reaction conditions. Among these are the Hok-

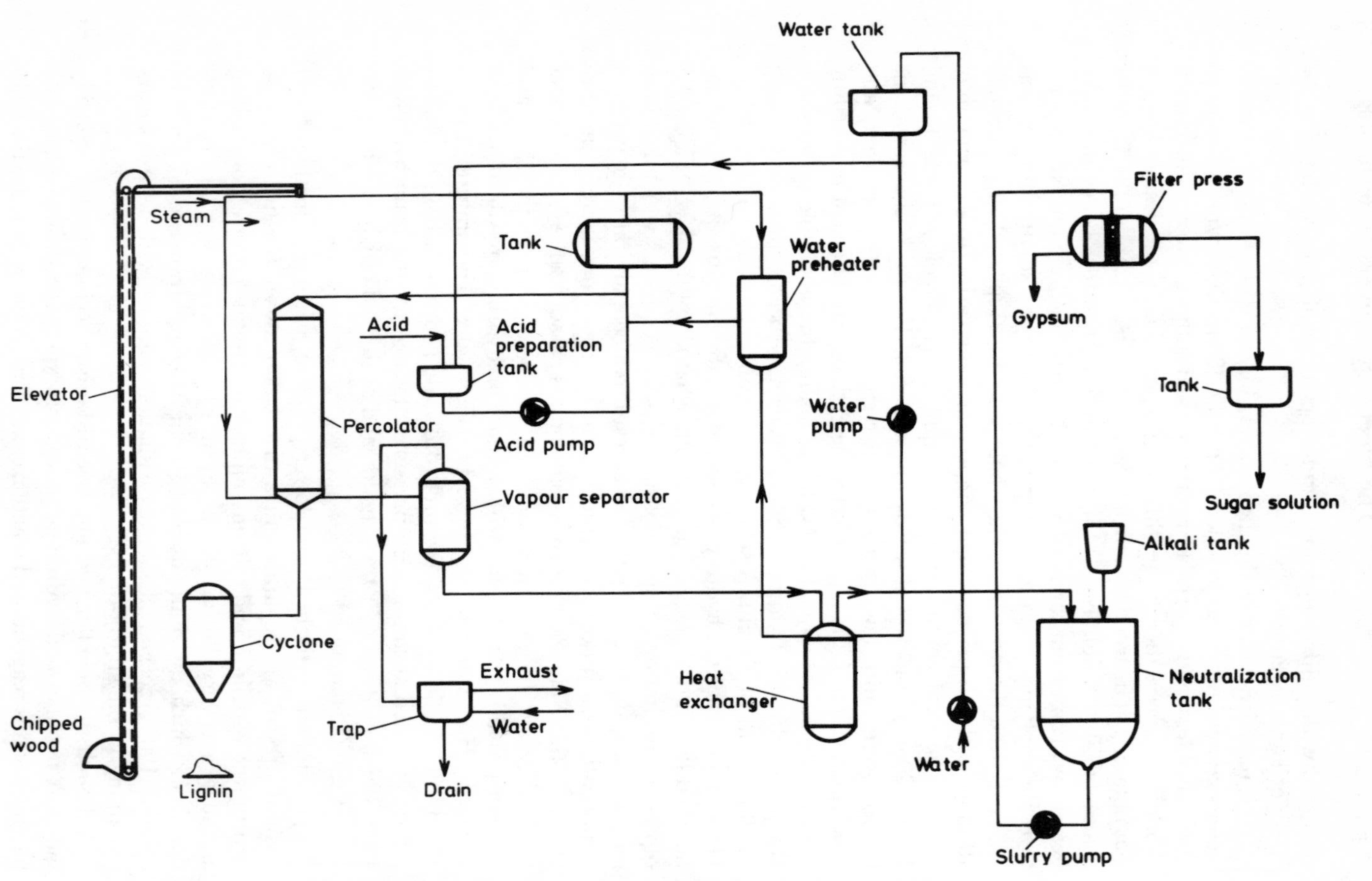

Fig. 4.6 Acid hydrolysis of wood – Scholler Process (after Oshima [101]).

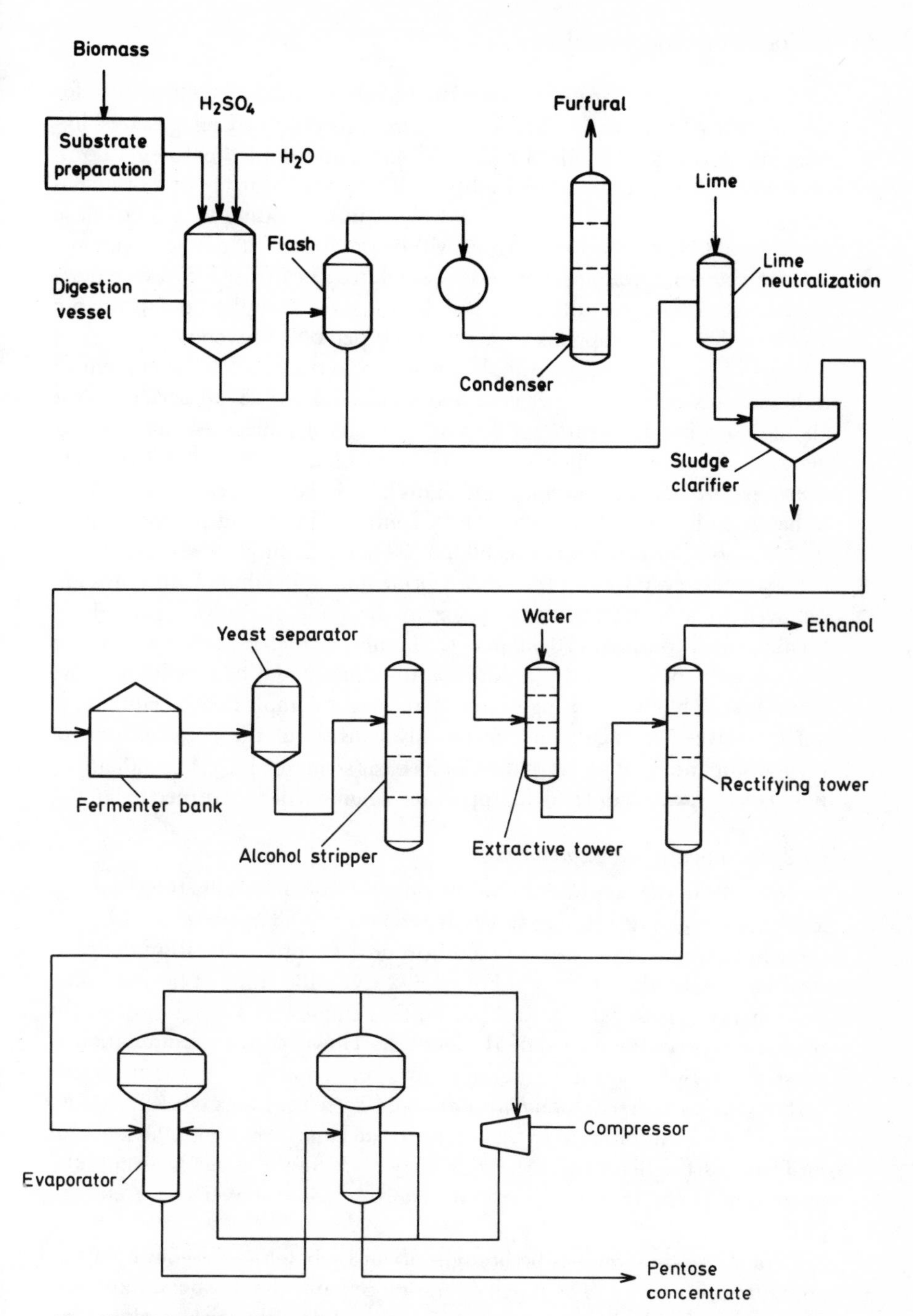

Fig. 4.7 Ethanol production via acid hydrolysis and yeast fermentation (after Bliss and Blake [52]).

kaido process using strong 80 per cent H_2SO_4 at ambient temperature for the main hydrolysis, the Noguchi–Chisso process utilizing gaseous hydrogen chloride in a fluidized bed at 40–45°C, and the Udic–Rheinau process with 41 per cent aqueous hydrochloric acid during the principal hydrolytic stage [52]. Enzymatic hydrolysis incorporating fungal cellulase enzymes as the saccharifying agent will be described later in the section.

Of the common cellulosic materials which may be utilized as raw materials for conversion to ethanol, wood is the most complex and probably provides the most problems as a result. 60 per cent of woody dry matter consists of potentially fermentable sugars on average, with the content of cellulose polymers 41–43 per cent and hemicelluloses 25–30 per cent. Not all the constituent sugars are hexose (6-carbon) molecules, some being pentoses (5-carbon) which cannot be fermented to ethanol by the *Saccharomyces cerevisiae* strains employed. This is especially the case with respect to hemicellulose polymers, which may contain 25 per cent pentoses (as in the softwoods, or conifers) or as high as 80 per cent (in hardwoods). These pentoses may even inhibit the hexose fermentation to ethanol, and thus are removed in a hydrolysis stage prior to the main hydrolysis procedure. Finally, wood comprises 20–30 per cent lignin, the higher percentages being characteristic of softwoods, in addition to cellulose and hemicellulose. The lignin invariably forms a high molecular weight complex with cellulose, is not hydrolyzed by acid, and chemically consists of the condensation of phenylpropane (C_6H_5–C_3H_7) units for its basic structure [52]. The difficulties posed by the presence of a high proportion of lignin will be returned to later.

4.3.2 Fermentation and distillation

Irrespective of the composition of the original raw materials, the ethanolic fermentation and distillation stages are common to all operations, although these in turn may have various modifications. Theoretically, from the classical equation $C_6H_{12}O_6 \rightarrow 2\ C_2H_5OH + 2\ CO_2$, 180 units of hexose sugar (principally glucose) should yield 92 units of ethanol on a weight for weight basis, i.e. a percentage yield of 51.1 per cent. However, the accumulation of yeast dry weight normally amounts to approximately 2 per cent of the carbohydrates utilized, along with another 2 per cent for glycerol production, 0.5 per cent for organic acids (chiefly succinic) and about 0.2 per cent for fusel oils (higher alcohols). Thus 47 per cent of the sugar is usually the maximum percentage by weight which is actually converted to ethanol [102].

Ethanol production can be brought about by batch, semi-continuous or continuous processes. The batch operation remains the most common, and it may reasonably be argued that, for a traditionally fairly simple production scheme yielding a product of comparatively low basic cost, the highly sophisticated operating systems often implied in continuous culture are economically superfluous. Nevertheless, all industrial grade ethanol derived from molasses in the USSR is currently manufactured by a modification of a continuous process, productivity may be increased by 1–2 per cent over the batch method, and the fermentation time required to give rise to a given quantity of product is halved. On the debit side, sterilization procedures must be very thorough since the consequences of contamination are greatly magnified.

Figure 4.8 is a flow diagram for a typical sugar cane blackstrap molasses batch fermentation plant. Yeast seed stages are sequentially passed through of volumes perhaps 100 litres, 1000 litres and 10 000 litres, with the fermentation stage of 100 000 litres capacity, although commercial size fermenters of 2 million litres are by no means unknown. The molasses medium itself is diluted to a wash containing fermentable sugars at around 12 per cent w/v concentration, which are fermented to ethanol at 6 per cent v/v within 36 hours. The temperature is maintained at 30–32°C, anaerobic conditions are rapidly set up with the evolution of CO_2, and the pH is maintained at pH 4–4.5. Unlike in the case of aerobic fermentations (e.g. in antibiotic production from moulds), the reactions are only slightly exothermic, and a substantial quantity of the heat emanating is lost from the fermenter by evaporation caused by the evolution of CO_2 gas saturated with water vapour. Normally in temperate regions sufficient cooling is effected by the outside circulation of water at ambient temperatures without the need for refrigeration. The medium itself may vary in composition, as where malt and grain wastes are utilized as the main substrate there is sufficient assimilable nitrogen, phosphorus, and magnesium present for maximal yeast growth and so no additional amounts are necessary, but with molasses and hydrolyzed cellulosic substrates, these often have to be added. In the particular case of molasses ammonium sulphate is often required to offset the N deficit.

Where continuous fermentation is used the *Saccharomyces cerevisiae* yeast strain is maintained at a reasonably high concentration, generally 7 g dry wt/litre or above, and the steady state sugar concentration is kept quite low, with ethanol leaving the fermenter at around 4.7 g/litre. Differences in steam requirements between batch and continuous culture are discussed in Section 5.5, and an energy analysis is presented in Table 5.7. By using alcohol-tolerant yeast strains bred to withstand higher sugar concentrations, and which ferment to around 12 per cent v/v ethanol [11], the increased ethanol concentration passing to the still allows for a 40 per cent reduction in distillation steam.

Prior to distillation the wash may be clarified by passage through a continuous separator or by partial settling of the cells and debris in the wash charger (as in Figure 4.8). Finally, in order to obtain satisfactorily pure ethanol, stills with highly efficient rectifying columns are used with

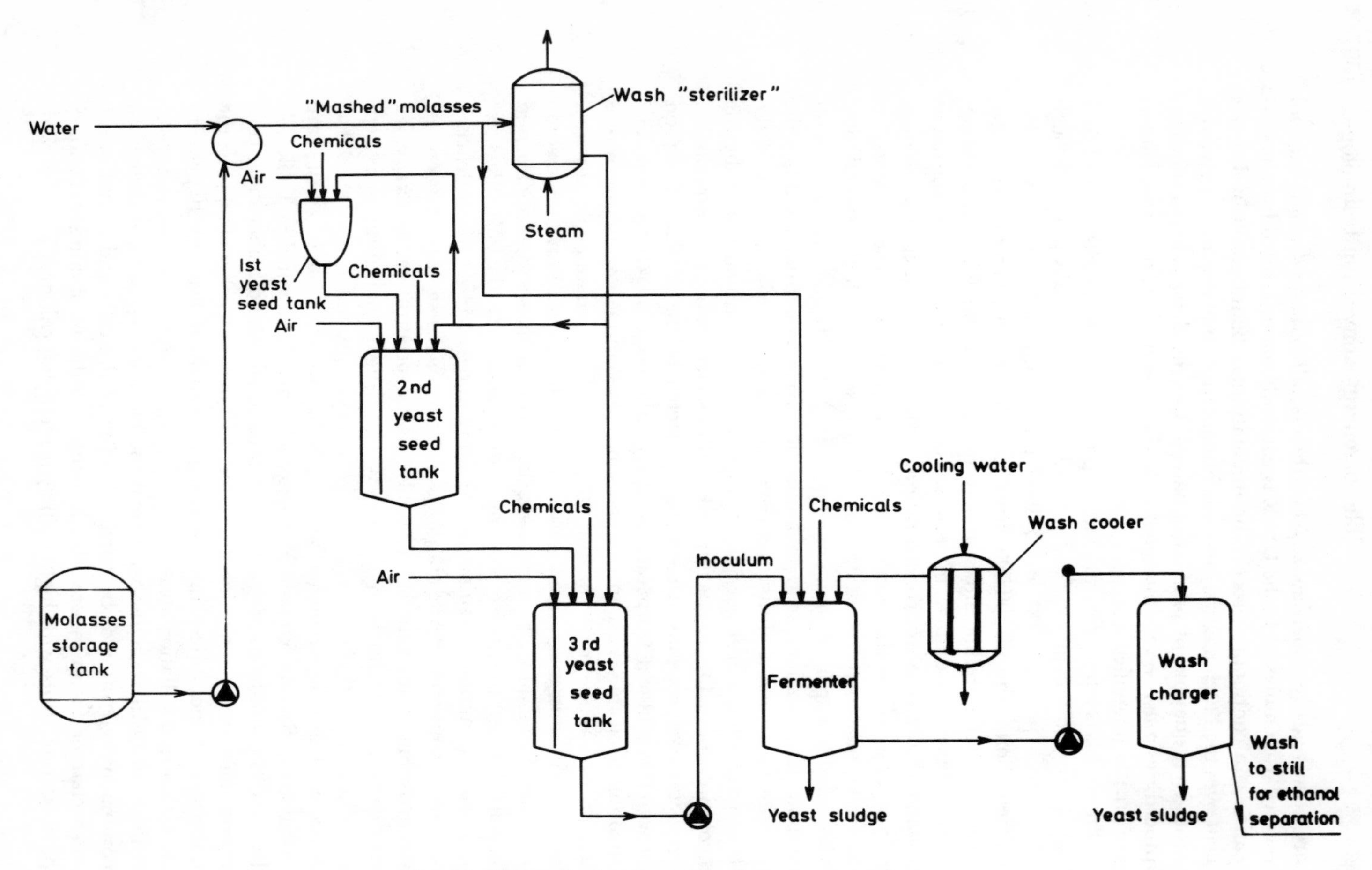

Fig. 4.8 Ethanolic fermentation of sugar cane molasses.

sufficient ability to concentrate the volatile impurities in relatively narrow zones of the vessels, and these can be drawn off as required. Fusel oils are concentrated near the base of the rectifying columns and may be removed without further treatment. Sometimes two stages of concentration are used, operating with pressure reduction between them. Ethanol concentration to around 95 per cent v/v occurs in a rectification tower, while the fusel oils are extracted in a separate vessel.

4.3.3 The biochemistry of ethanol production

The provision of a biochemical, as well as a process, description of the events leading to the metabolism of ethanol from carbohydrates enables the tracing of energy flow from the sun to its capture by plants on earth (see Chapter 2) to be continued on to a form in which it can be readily utilized by man. The energy-rich molecules of adenosine triphosphate (ATP) and the oxidation-reduction reactions involving the nicotinamide adenine nucleotide cofactors and their reduced, hydrogenated forms are again much in evidence, as they were in the photosynthetic mechanisms described in Section 2.2. Such molecules are basic to the great majority of living organisms, whether plants, animals, fungi, or bacteria. What is more important is that all living systems must obey the laws of thermodynamics. The second law, for instance, states that in any energy-yielding reaction the total energy liberated is never totally available for the performance of work, as some of it is lost as an increase of entropy. However, the actual amount of energy derived from biological energy-generating or degradative reaction sequences (catabolism) that can then be applied to energy-consuming processes is always significantly less than the maximal available (or free energy potential), even after accounting for the constraints imposed by thermodynamics itself. Thus, in the present case, concerning the alcoholic fermentation of sugars by yeast, the free energy evolved by the formation of two moles of ethanol and two of CO_2 from one carbohydrate mole is, according to the equation

$$C_6H_{12}O_6 \rightarrow 2\ CO_2 + 2\ C_2H_5OH,\ 234.5 \times 10^3 J.$$

In fact only around 67×10^3J of this energy is available, making an overall thermodynamic efficiency of 29 per cent, since the remainder is dissipated as heat. This 67×10^3J of retained energy is equal to the energy bound up in the two terminal phosphate bonds of a mole of the energy-carrying ATP. On hydrolysis of one of these bonds a terminal phosphate group may be transferred to an acceptor molecule, with the bond energy present being almost wholly preserved.

A common and vitally important example is afforded by the activation of glucose:

$$\text{glucose} + \text{ATP} \rightarrow \text{glucose} - 6 - \text{phosphate} + \text{ADP}$$

This reaction is particularly significant since it initiates the fermentation of glucose to pyruvic acid via the glycolytic Embden–Meyerhof–Parnas (EMP) pathway, virtually ubiquitous throughout the living world. Several steps, all enzymatically catalyzed, are sequentially involved, and the pyruvic acid formed may then undergo various reactions to produce, for example, lactic acid in animal tissues, some fungi, protozoa, and bacteria; acetic, succininc, formic, propionic, and butyric acids in different bacterial groups as well as acetone, butanol, isopropanol, and 2,3- butanediol; and finally ethanol in plant tissues and certain fungi. It is the last of these that one is concerned with here, namely the alcoholic fermentation by *Saccharomyces cerevisiae* of hexose sugars, chiefly glucose. The biochemistry involved in converting cellulose, starch, etc. to fermentable sugars is discussed briefly in this section, but since the book is primarily one dealing with energy rather than biochemistry many of the finer details have been omitted though they can of course be found easily in the relevant biochemical literature [103, 104].

Figure 4.9 divides the conversion of glucose to ethanol via the EMP pathway into four distinct stages. It should be stressed that the overall process operates entirely under anaerobic conditions. From the original glucose molecule fructose –1.6-diphosphate is formed via phosphorylation reactions at the expense of two ATP molecules in the initial stage. Cleavage of the C_6 diphosphate results in the production of two C_3 phosphates, dihydroxyacetone phosphate and glyceraldehyde phosphate, which themselves are readily interconvertible. One molecule of 1,3-diphosphoglyceric acid is obtained from each C_3 compound via the oxidation of the triose phosphate and a reduction of NAD, accompanied by an esterification of inorganic phosphate. Two molecules of 1.3- diphosphoglyceric acid formed in stage 2 are then converted via intermediates to two molecules of pyruvic acid. In each case, both phosphate groups are transferred to ADP so that a total of four moles of ATP are produced per mole of glucose utilized. However, as two moles of ATP were previously expended in the initial activation reactions, the net ATP yield is in fact two moles. The overall reaction thus far may be represented thus:

$$\text{Glucose} + 2\ \text{Pi} + 2\ \text{NAD} \rightarrow 2\ \text{pyruvic acid} + 2\ \text{ATP} + 2\ \text{NADH}_2$$

Stage 4 sees the culmination of the reaction sequence in which the pyruvic acid is decarboxylated to CO_2 and acetaldehyde, the latter being reduced to ethanol while $NADH_2$ is simultaneously reoxidized. Thus, combining the maximum attainable photosynthetic efficiency of around 5.3 per cent, as

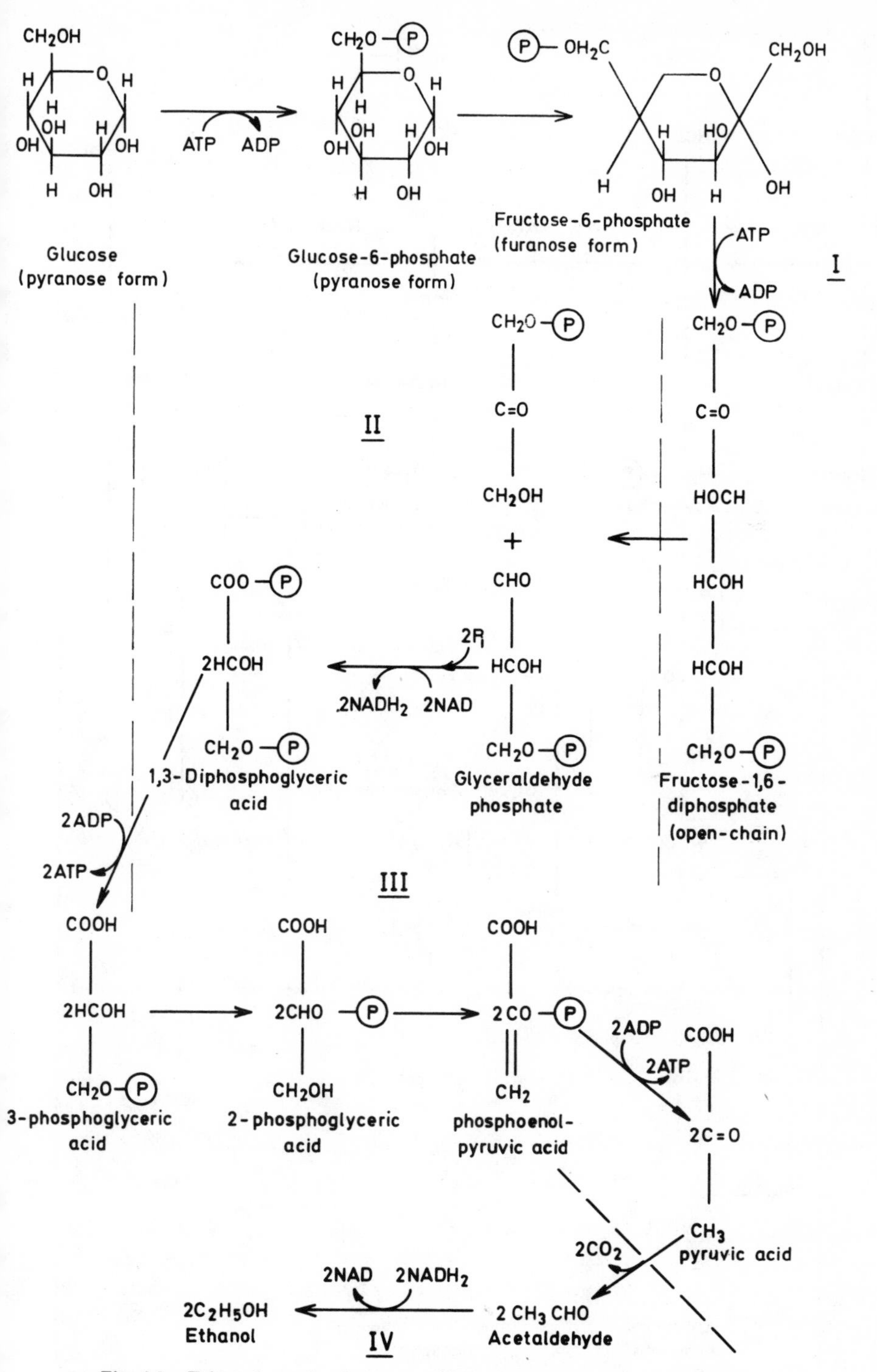

Fig. 4.9 Ethanol production route from glucose via the EMP pathway.

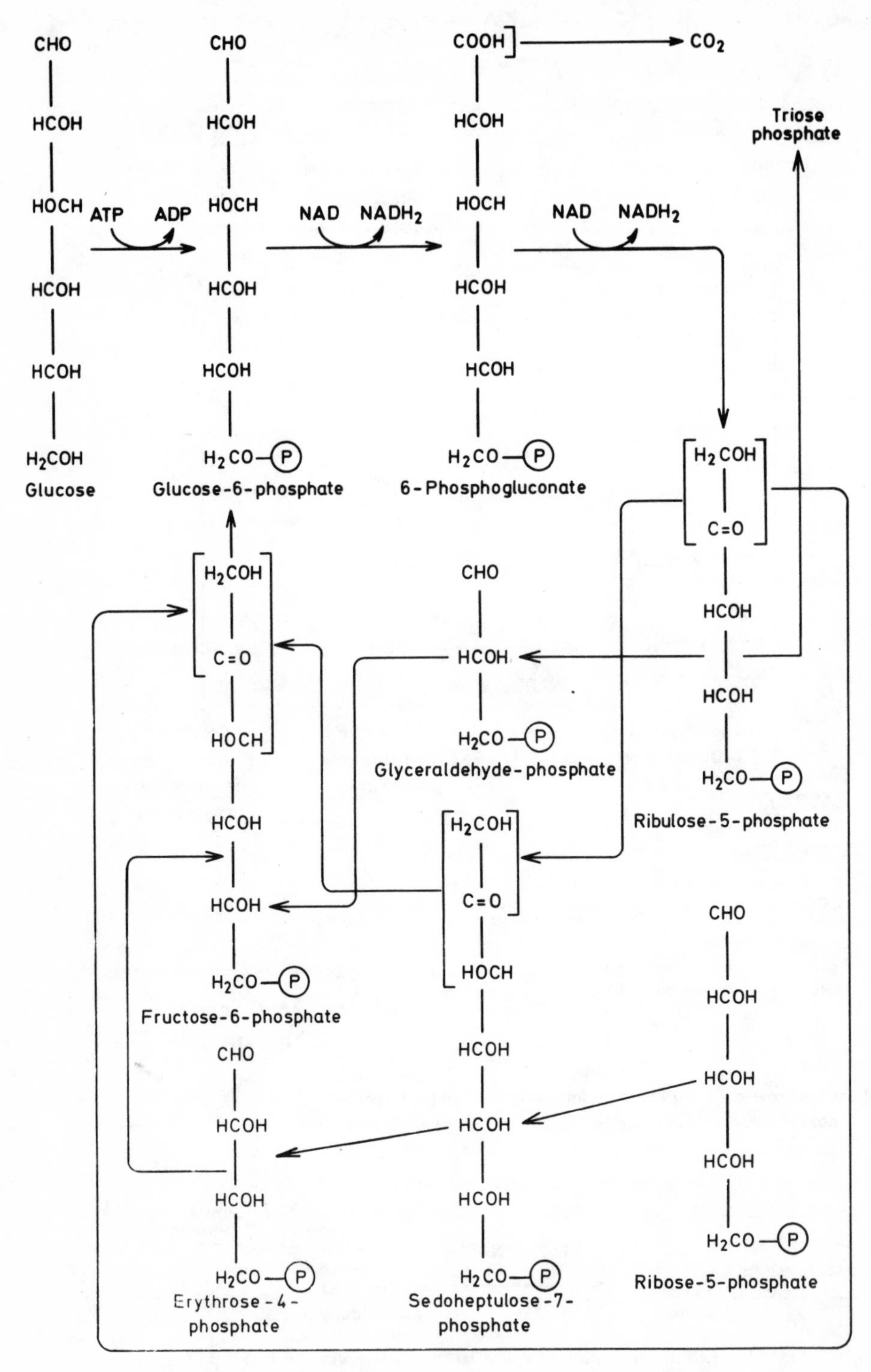

Fig. 4.10 The pentose phosphate pathway from glucose.

stated in Section 1.2, for the incorporation of solar energy into plants as carbohydrate energy, with the 29 per cent efficiency of carbohydrate fermentation to ethanol described above, an overall efficiency of solar energy conversion to ethanol of 1.5 per cent is realized. The practical efficiency would obviously be considerably less in a necessarily imperfect system.

Although the EMP pathway is the most common route for the production of pyruvic acid, and thence ethanol, from sugar substrates by yeasts, other pathways may also play a supporting role, chief among which is the so-called pentose phosphate (PP) pathway, illustrated in part by Figure 4.10. Glucose-6-phosphate is the crucial compound here, initially degraded via 6-phosphogluconate to the pentose, ribulose-5-phosphate and CO_2 [105].

Subsequently a series of intermediate molecules containing both phosphate and a range of carbon atoms numbering from two through to seven may arise, before triose phosphates like dihydroxyacetone phosphate and glyceraldehyde phosphate are formed (Figure 4.10). The remaining reactions of the EMP pathway then come into play. The PP pathway is an example of a hexose monophosphate shunt, whereby the reaction mechanisms leading to the production of C_3 intermediates are different to those in the complete EMP pathway, with a result in energy terms indicating that the sequence is less favourable. It does have biosynthetic significance (for instance the intermediate ribose-5-phosphate is an essential component of the genetically vital RNA), but when glucose is fermented through a hexose monophosphate pathway it should be observed that the net yield of ATP is one mole per mole of glucose converted to pyruvate, and thence to ethanol, i.e. only 50 per cent that usual in the EMP pathway.

Thus the energy flow throughout this particular conversion route has been outlined here, although not of course in its minutest biochemical detail. It is less easy to trace the chemical reactions involved in some of the conversion methods such as pyrolysis, where a multiplicity of end-products may be formed, but mention will be made of these as they occur in the text.

4.3.4 Ethanol from starch

Intermediate between monosaccharide and cellulosic molecules with respect to the extent of pre-treatment required prior to ethanolic fermentation come the substrates based on starch configurations. Starch is the principal storage carbohydrate of higher plants and consists of two components, amylose and amylopectin, which are present in varying amounts. Amylose consists of glucose units joined linearly by α–1–4 linkages (the numerals each relate to a specific carbon atom in the molecular structure), and has a molecular weight of from less than 10 000 up to 150 000. Amylopectin is branched, with amylose-type units consisting of around 30 glucose molecules being additionally linked to each other via α–1–6 linkages, and molecular weights of 500 000 or more are common [103].

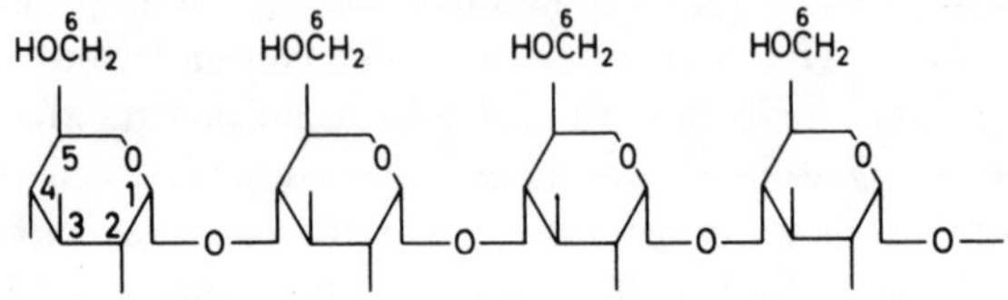

Amylose

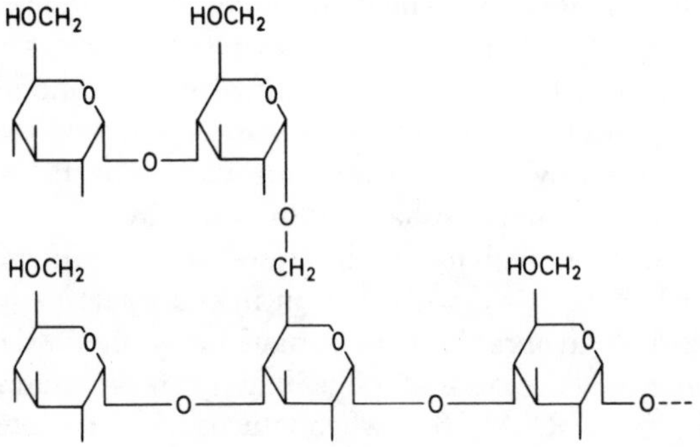

Amylopectin

Of the plants with a high starch content which are candidates for cultivation as an 'energy crop' probably cassava is the most promising. Cassava is native to Brazil and other South American countries, but has also been introduced into many other tropical countries. Its tubers contain 25–32 per cent starch (wet wt) which can be dried and cut into chips containing about 75 per cent starch and 7 per cent moisture. In the context of the Brazilian national alcohol programme mentioned elsewhere, the plant offers an inexpensive source of starch, but requires a suitable saccharifying agent for its initial conversion to fermentable sugars.

Starch hydrolysis by mineral acids at pH 1.5–2 under 2.1 kg/cm^2 steam pressure for 20–30 min has been practised commercially, but has the inherent disadvantages of equipment corrosion, low yields (unless nutrient supplements are added to the ensuing fermentation liquor), and low recovery and high salt content of fermentation by-products. Barley malt could be used for its enzymatic action, but is scarce and costly in Brazil, thus increasing the overall ethanol production cost. Corn malt is also unsatisfactory due to the difficulty in obtaining good malt under tropical conditions; bacterial contamination frequently results from the indigenous high temperature and humidity, and corn malt is, in any event, an inefficient conversion agent.

Alcohol yields of 43–74 per cent of the theoretical maximum have been reported following acid hydrolysis; conversion with barley malt has resulted in yields of 70–74 per cent of maximum; while equivalent amounts of corn malt have proved even less productive. However, when mould bran enzyme preparations have been employed yields of 80–85 per cent have been obtained, and it has been demonstrated possible to obtain overall plant efficiencies of 90 per cent from cassava mashes acted upon by submerged fungal cultures [17]. These differences in efficiency may be attributed to the molecular configuration of cassava starch, which is more amenable to amylase enzyme attack where the reacting enzyme complex (including α–amylase, amyloglucosidase, and limit dextrinase) is of fungal origin. Where fungal amylases are produced commercially by submerged-culture methods for the conversion of starchy raw materials in the production of industrial alcohol, then the enzymes are almost exclusively utilized without previous concentration and purification [106].

In order to render the starch more responsive to amylase attack the cassava tubers are initially ground and autoclaved in an aqueous suspension, after which the starch shows increased solubility.

The enzyme production stage begins with spore inoculations of a suitable amylase-secreting strain of the ascomycete fungus, *Aspergillus niger* from solid laboratory cultures into flasks of liquid medium. These flasks are incubated at 30°C for three days on a reciprocal shaker, the contents transferred to an aerated and agitated seed tank for 24 hours, and finally the inoculum is passed into the stainless steel production vessel containing possibly 40 000 litres of medium. The medium itself contains soluble starch to induce enzyme production, as well as cornsteep liquor as a nutrient source of nitrogen, and other key elements. These materials may both be obtained from the growth of corn, but where this crop is not or cannot be cultivated alternative sources are cassava itself and possibly distillers' solubles respectively.

The batch fermentation is allowed to proceed at 28°C for seven days with aeration and agitation at pH 4.5. At the termination of this culture period the fungal mycelium will be quite extensive, and the quantity of α–amylase accumulated in the medium attains its maximum. The individual components of the extracellular enzyme complex are often described by a variety of names. Of these, amyloglucosidase is the main saccharifying agent, although the culture broth will also possess dextrinogenic activity. Enzyme yields are expressed in terms of saccharogenic and dextrinogenic activity units/ml of culture. One unit is defined as the amount which catalyses the transformation of one micromole of substrate/min under the operating conditions of starch hydrolysis, these to be outlined in the following paragraph. Expected yields would be 9920 units/ml of dextrinogenic activity and 158 units/ml of saccharogenic activity, the latter indicating the potential of starch conversion to glucose. 6.32×10^6 saccharogenic units would constitute 1 per cent of a single production line's output. Finally, the fermentation broth is filtered to remove the mycelium, which is a potential source of protein and could be used as a livestock feed, or even recycled to reduce the quantities of necessary added nutrients [107, 108].

The enzymatic hydrolysis reaction is accomplished at pH 4.5 and temperature 50 °C within an agitated vessel for 60 hours, when 90 per cent of the starch substrate can be recovered as glucose. The original starch suspension should have at least a 30 per cent solids content to make the process economically viable, and so a preliminary mild acid hydrolysis is usually required to reduce the viscosity [109]. 1.89×10^6 saccharogenic enzyme units are required to produce 1 t of glucose, i.e. 0.3 per cent of the output from one enzyme production line. A conceivable energy analysis of the whole process is presented as Table 5.8, including the inputs to the above stage plus the alcoholic fermentation and distillation operations already discussed.

4.3.5 Ethanol from ligno-cellulose via enzymatic hydrolysis

To complete this section on ethanol production it is necessary to deal with the enzymatic hydrolysis of cellulosic or ligno-cellulosic materials to sugars, and the subsequent fermentation and distillation procedures. The overall process incorporates features of all the previous technologies, but at present is the most unattractive from a net energy analysis viewpoint, as the discussion in Section 5.5 will indicate. The ligno-cellulosic complex outlined

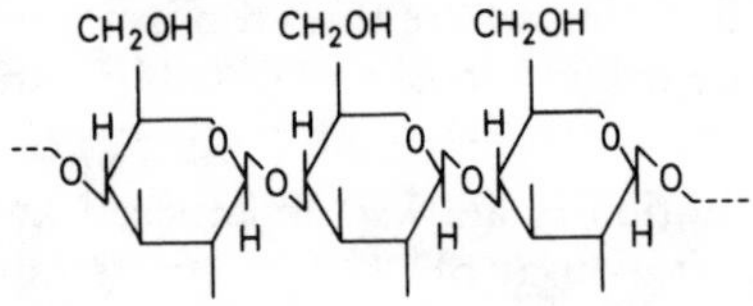

Cellulose

above is very resistant to rapid degradation by microbial enzymes, not only due to the protective effect of the three-dimensional lignin polymer, but also due to the presence of β–1–4 linkages in the cellulose polymer as opposed to the more readily hydrolysed α–1–4 linkages of starch. Only the wood-rotting fungi are capable of splitting the ligno-cellulose complex into lignin breakdown products and native cellulose (see Section 7.6), and this at a very slow rate. The alternative of acid hydrolysis has been described earlier, but with the inevitable disadvantage of the requirement for expensive corrosion-proof capital equipment, the possible occurrence of sugar decomposition, and the formation of unwanted by-products by reaction of the acid with impurities which are invariably associated with waste cellulosic substances.

Glucose syrups produced enzymatically tend to be more pure and constant in composition; moreover, the enzyme is specific for cellulose and the reaction takes place under moderate conditions, resulting in higher glucose yields than generally arise following acid hydrolysis.

There are four major stages involved in the microbial conversion of cellulosic materials to ethanol, as in the starch to ethanol route described above. These entail:

1. Treating the raw material to render it susceptible to enzymatic attack
2. Producing the necessary enzyme(s) (in this case cellulase)
3. Hydrolysing the prepared substrate to sugars (chiefly glucose)
4. Fermenting the sugars formed to, and distilling off, ethanol.

In order to release the cellulose from its binding lignocellulosic complex either acidic or alkaline hydrolysis or mechanical grinding is required, with ball-milling being easily the most effective pre-treatment for decreasing particle size and reducing the cellulose's crystallinity. However, the energy input to the ball-milling procedure is considerable (see Section 5.5), and makes the economics of the whole process unfavourable at present. Nevertheless, the resistance of ligno-celluose to degradation does vary from one material to another of different origin, and this will have some influence on the extent of treatment needed.

The scheme for the degradation process is outlined in Figure 4.11. On completion of reaction L by non-microbial methods the native cellulose formed becomes amenable to attack by a wider range of fungi and bacteria, but still at a comparatively slow rate. In practice the above treatments will reduce the native cellulose to the reactive form via C_1. At this stage the compound is susceptible to degradation by the cellulase activity of several types of organism, which may conveniently be categorized according to oxygen and temperature requirements as aerobic mesophiles, anaerobic mesophiles, aerobic thermophiles, and anaerobic thermophiles, with the product being, in the main, monosaccharide sugars.

Where ball-milling is selected as the initial treatment using, for example, dry newsprint as the waste cellulosic raw material, researchers at the US Army Natick Development Center in Massachusetts have needed an electricity input of 1.64 kWh/kg of newsprint in order to obtain a mean particle diameter of 50 μm. At least 50 per cent of this raw material will be saccharified via cellulase hydrolysis as a result [111–113]. With cellulose constituting 61 per cent of waste paper on average, and at a conversion efficiency to sugars reported at 82 per cent, 1.8 kg (dry wt) ball-milled newsprint will yield 1 kg of sugars at an electrical requirement of 2.95 kWh. More difficult substances such as wood chips and sawdust would additionally require hammer-milling and screening prior to ball-milling, and thus consume even greater quantities of electricity before assuming a state suitable for hyrolysis. It has been suggested that a much less energy-intensive operation involving hammer-milling alone is sufficient to reduce 80 per cent of newsprint to a mean particle size diameter of 70 μm, and that this is adequate for subsequent satisfactory enzymatic hydrolysis, though apparently as yet this theory has been unsubstantiated in practice [114].

Probably the most promising conversion of reactive cellulose to cellobiose and then to glucose is that mediated by enzyme preparations from the aerobic, mesophilic, cellulolytic mould, *Trichoderma viride* [115]. The most

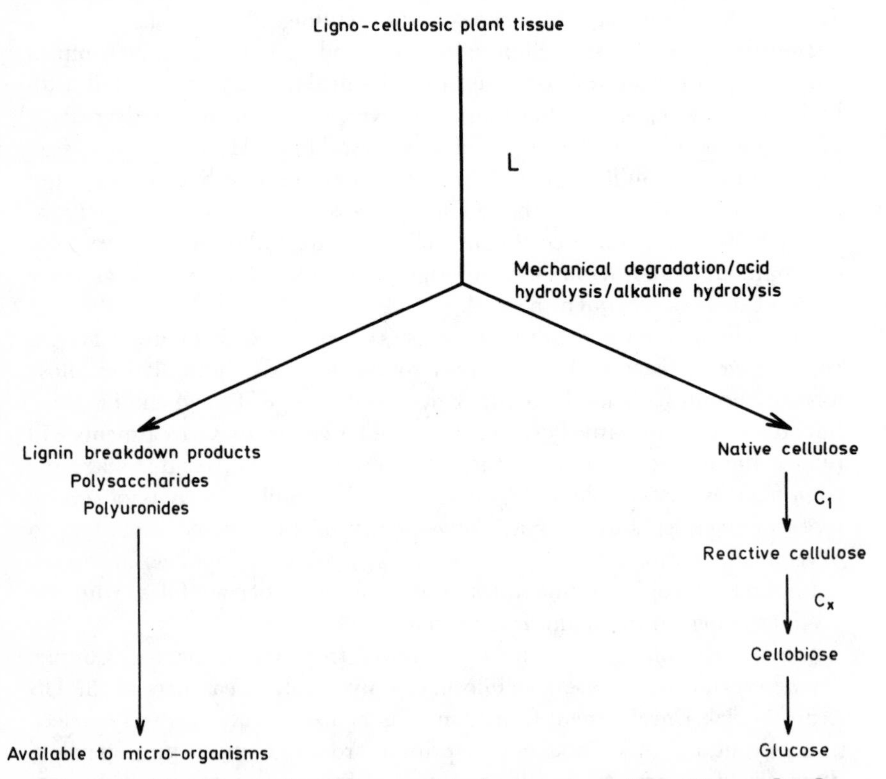

Fig. 4.11 The degradation of ligno-cellulosic plant tissue (after Worgan [110]).

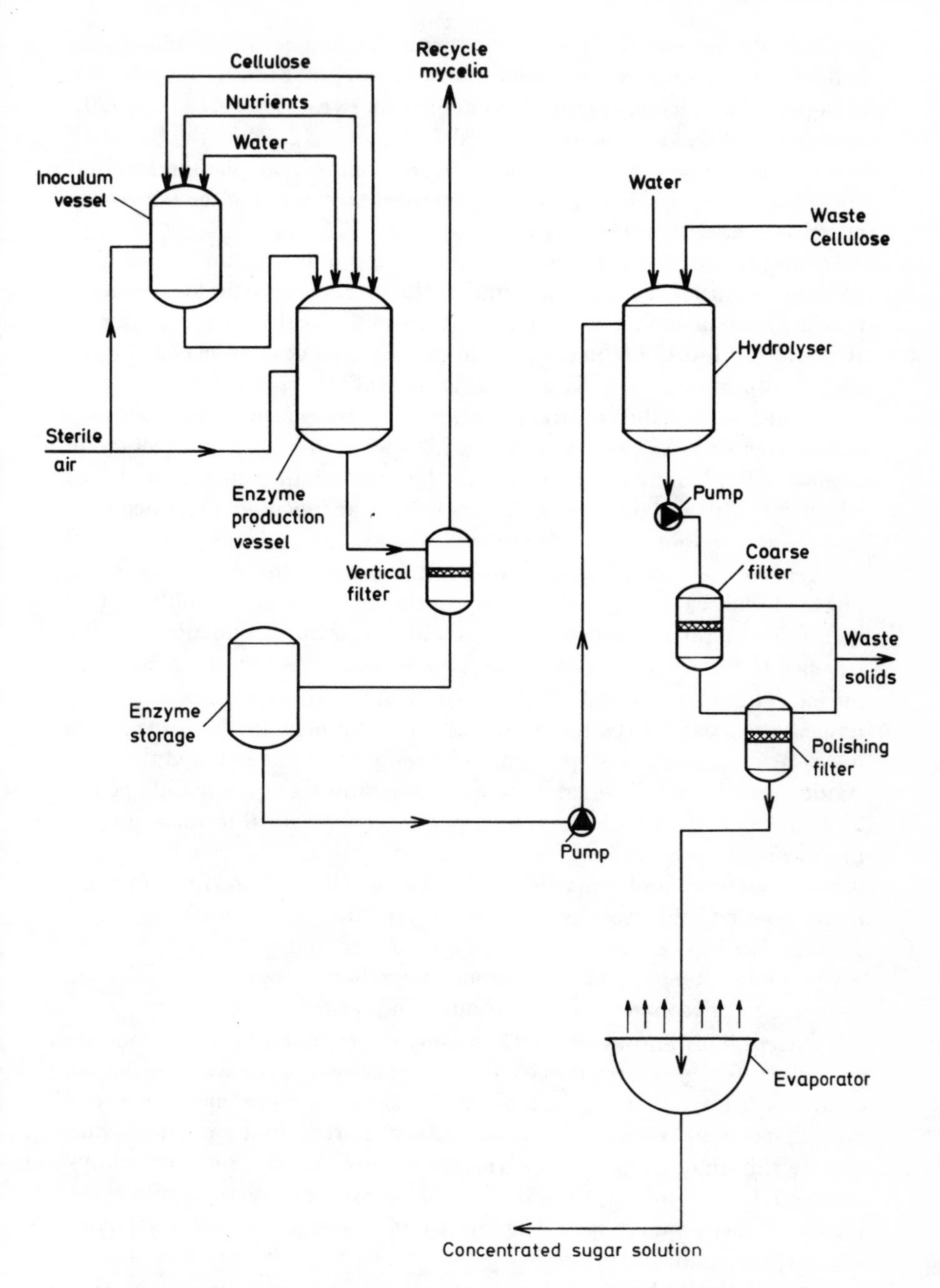

Fig. 4.12 Simplified cellulose to sugar enzymatic conversion plant.

in-depth studies have been performed with this fungus, chiefly employing ball-milled newsprint as the cellulosic substrate, at Fort Natick where a pilot-plant has been in operation for at least four years [111–113, 116–120]. A diagram of this scheme is presented as Figure 4.12, while the following description is that envisaged for a future commercial plant under full operating conditions, based primarily on work carried out at the University of California, Berkeley [114].

In this production process an inoculum of the mutated strain *T. viride* QM 9414 is introduced at a v/v concentration of 10 per cent into the fermentation vessels. This inoculum is prepared at 29–30°C, with uncontrolled pH, aeration at 0.1–0.15 v/v/min, a minimum dissolved oxygen level at 15 per cent of saturation, and a K_1 a (max) of 60 mM?$_2$/1–hr–101 kPa.

Conditions for cellulase production are such that a semi-continuous process is operated. Vessel 1 functions at 80 per cent of its total volume of around 380 m^3 until the fermentation has passed through 25 hours, at which point 10 per cent of the culture is used as an inoculum to fermenter 2. In turn the second vessel is operated for 25 hours and then transfers 10 per cent of its culture volume to the third tank, the procedure being repeated with each succeeding vessel in the line. Vessel 1 completes a 90 hour fermentation, considered as optimum for enzyme production, and the product is then harvested and the tank sterilized before receiving fresh media. This turn around is estimated to take 10 hours, giving a total fermentation time of 100 hours, and at the 100th hour the fourth vessel in line is ready to transfer 10 per cent of its contents to the first, and the cycle continues as before. The fermentation temperature is again maintained at 29°C ± 1°C and with the remaining parameters identical to those used in the inoculum preparation, but with a reduced aeration requirement of 0.05–0.1 v/v/min. and a maximum impeller speed of 220 rev/min (a maximum power to the broth of 0.52 kW/m^3 resulting). Medium constituents include shredded cellulose for enzyme induction and carbon source, with ammonium sulphate, urea, and peptone providing nitrogen, and potassium dihydrogen phosphate acting as a source of phosphorus.

Sufficient lines to provide 10 000 m^3/day of broth at an enzyme concentration of 1.03 kg/m^3 are conceived, and 90 per cent of the water is evaporated off to leave a concentration of 10.3 kg/m^3, the enzyme then being precipitated with acetone, refrigerated, and stored. In terms of enzyme activity this amounts to 10^{10} International units per day, sufficient to hydrolyse 770 t ball-milled newsprint to 430 t sugars. For every kg of enzyme produced there would be 3.2 kg dry wt of harvested fungal mycelium, which could again be recycled or possibly used as a livestock feed.

The hydrolysis of the cellulose substrate by the enzyme is envisaged to occur in five agitated cylindrical concrete digesters of the type used for solid

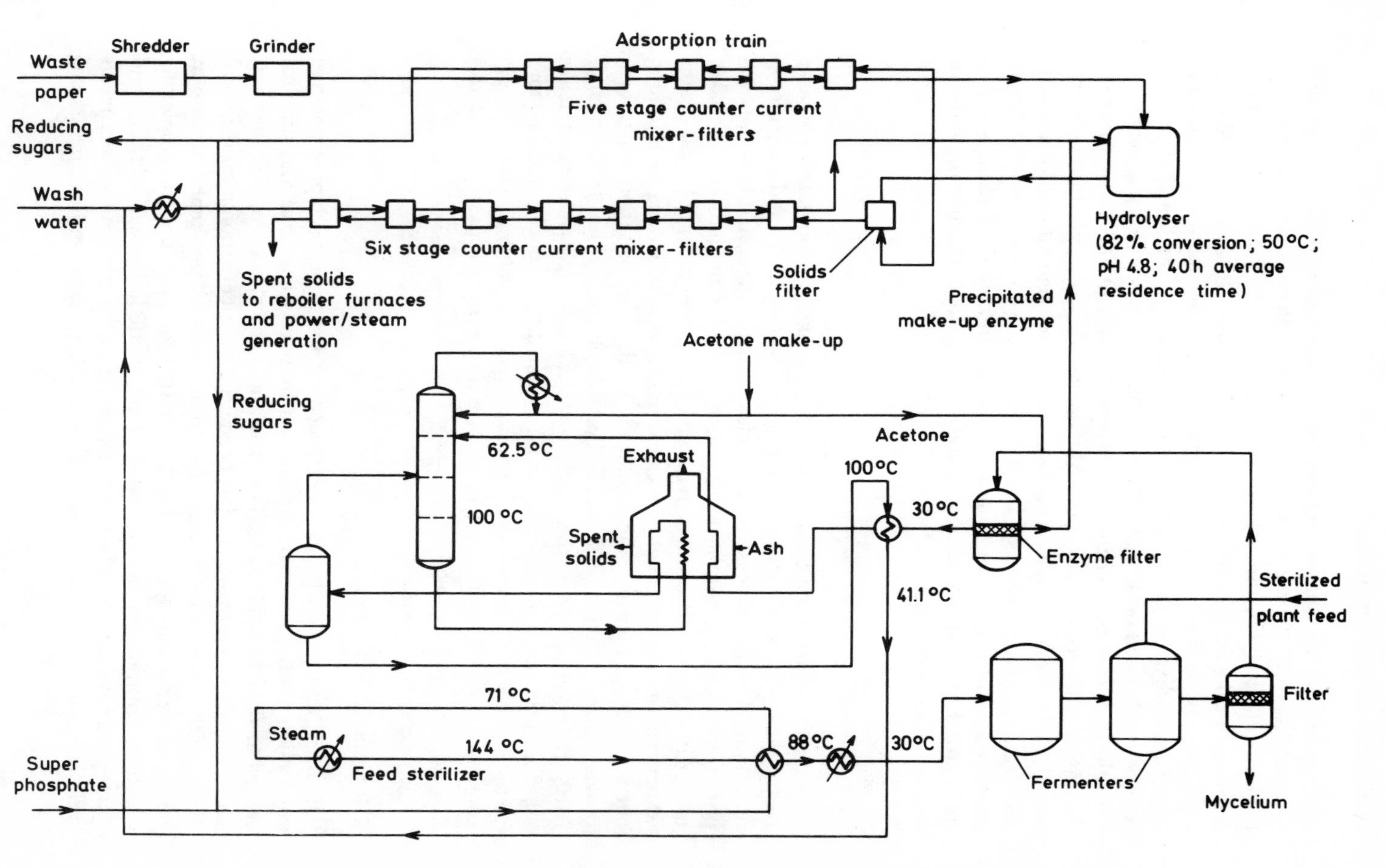

Fig. 4.13 Cellulose to sugar enzymatic hydrolysis plant (after Wilke and Mitra [114]).

waste treatment in sanitary engineering. The average reaction time at 50°C and pH 4.8 is 40 hours, realizing a cellulose conversion of 82 per cent. Following hydrolysis, the effluent is filtered in a vacuum drum filter to separate out the liquid and solid phases. The liquid stream then enters the adsorption train, corresponding to a mixing tank and vacuum drum filter at each stage, whereby the enzyme is recovered within the liquid phase and returned, adsorbed on to the cellulosic solids, to the hydrolysis vessels. A realistically recoverable enzyme activity of 20 per cent has been postulated. The spent solids, containing 1 kg H_2O/kg dry cake are utilized as the source of fuel for the boiler-furnace and for steam/power generation. Meanwhile, the sugar product passes to the alcoholic fermentation and distillation stages described previously. A flow diagram of the conceived plant is shown in Figure 4.13, with an energy analysis using wood as the raw material source to be found later in Table 5.10 [114].

4.3.6 Reducing the energy requirements

This then completes the production process range of ethanol from organic materials, covering a comprehensive range in complexity and difficulty of operation. The basic yeast fermentation and distillation stages are well proven technologies, but the energy requirement for distillation is probably too great at present. It could be reduced by the development of yeast strains which tolerate high alcohol concentrations, and by selecting or mutating out satisfactory thermophilic organisms so that fermentation waste heat could be used to actually distil off the ethanol from within the fermenter itself. However, perhaps the single most critical factor needed to make ethanolic fermentations more attractive, particularly in non-tropical regions of the world, is a less energy-intensive process involving cellulosic substances. Promising, but as yet incomplete progress has been made in this general area, and this will be discussed later in Chapter 7.

4.4 Chemical reduction

The final, and presently least developed, of the aqueous biomass-to-fuel conversion methods is that of reduction by purely chemical and physical means to yield valuable fuel oil or substitute natural gas. An alkaline catalyst such as sodium carbonate has been used to react carbon monoxide and steam with a waste cellulosic slurry containing 85 per cent moisture at temperatures of 250–400°C, 13.8–27.6 × $10^6 N/m^2$ pressure, and employing vigorous agitation, to yield an oil of approximate formula $(C_{11}H_{19}O)_n$ and energy content nearly 40 MJ/kg. Upwards of 23 per cent of the feed's cellulose content is thus converted, but with process improvements it is anticipated that the theoretical maximum efficiency of around 50 per cent will soon be approached. The chemistry of the reactions has yet to be fully elucidated, but under the prevalent conditions hydrogen is certainly formed, as follows:

$$CO + H_2O \rightarrow CO_2 + H_2$$

Oxygen is thus progressively removed via gas-liquid separation, with the simultaneous reduction of the carbohydrate to aliphatic hydrocarbon, which may also retain a certain proportion of oxygen if the reaction remains incomplete. Nevertheless, this reduction appears to be facilitated more by the presence of carbon monoxide than by hydrogen, and so the net conversion is probably roughly intermediate between the following two reactions:

$$C_6H_{10}O_5 + CO \rightarrow C_6H_{10}O_4 + CO_2$$

$$C_6H_{10}O_5 + 5\,CO \rightarrow C_6H_{10} + 5\,CO_2$$

Evidence supporting this reasoning is afforded by the reactions involving the alkaline catalyst, which takes part in the conversion as formate, thus:

$$NaHCO_3 + CO \rightarrow NaOOCH + CO_2$$

$$NaOOCH + C_6H_{10}O_5 \rightarrow C_6H_{10}O_4 \text{ (oil)} + NaHCO_3$$

Direct reduction of the cellulose by the formate occurs, which then reverts to bicarbonate, this in turn reacting with fresh carbon monoxide to reform formate.

When the reaction temperature is approximately doubled and the pressure reduced by 30–50 per cent a high-energy gas is the principal end-product. It appears that the above-mentioned formate in this case decomposes to evolve an active hydrogen and to regenerate the bicarbonate:

$$NaOOCH + HOH \rightarrow NaHCO_3 + \text{'}H_2\text{'}$$

The oil which would have been produced is hydrogenated by the gasified active hydrogen to form instead gaseous alkanes.

Consequently, greater effort appears to be concentrated on oil production, particularly from pulverized, partially dried wood chips, where problems of feeding solids into a pressurized reactor are being overcome either by introducing a wood-oil slurry or by pre-pressurizing the feed system with process carbon monoxide. A major virtue in the case of a woody, raw material is that the lignin, as well as the cellulose component, is amenable for conversion into oil, and also even toxic inputs are acceptable; particularly important with respect to utilizing wastes [52].

4.5 Gasification

Dry chemical conversion processes may be broadly divided into heating the biomass in the absence of air, known as pyrolysis (dealt with in Section 4.6), and heating the biomass in the presence of limited quantities of oxygen for the maximum liberation of carbon monoxide and hydrogen (synthesis gas), known simply as gasification. A third category, hydrogasification is known whereby dry cattle manure and cellulosic substances have been reacted with hydrogen at 540°C and $6.9 \times 10^6 N/m^2$ to produce methane and ethane, but its prospects appear limited owing to the fact that hydrogen itself is a valuable fuel, and its use in such a process may not be justified on energetic grounds alone [27].

Gasification, however, certainly is a developing technology, particularly in the fields of converting coal and municipal refuse to a gaseous product. In addition to these raw materials, wood is also being utilized as a substance suitable for gasifying, the crude gas produced consisting of varying amounts of carbon monoxide, carbon dioxide, and hydrogen, and, depending on the type of process employed, perhaps large quantities of nitrogen. Small amounts of methane and heavier hydrocarbons are also formed, approximately 2 per cent dry wt of wood being converted to a recyclable oil-tar fraction.

4.5.1 Low-energy gases

The partial oxidation procedure may be performed utilizing either oxygen or air, at atmospheric pressure in both instances. A crucial difference between these two approaches is that around 42 per cent of nitrogen is contained in the ensuing gaseous mixture when air is used, resulting in a lowering of the product's heating value, occasionally by fully 50 per cent; 7 MJ/m^3 as against up to 14 MJ/m^3 on average. Also, a high proportion of moisture is evident, which may easily be removed in the subsequent upgrading of the gas quality. It is envisaged that the gaseous product, which is a low-medium energy containing fuel, will be subsequently converted to a more useful substance such as substitute natural gas (SNG), methanol, or ammonia. In the manufacturing of these valuable commodities the input gases are preferably required at $2.1–2.8 \times 10^6 Nm^{-2}$ to enable yield maximization to be effected; and so under these reasonably high pressures the absence of inert nitrogen constitutes a distinct advantage. Therefore, gasification of biomass with oxygen rather than air is probably the more desirable. The following is a general description of the US Purox process, which has been used for municipal solid waste (Figure 4.14), and which could be extended to cater for other potential organic materials, including wood chips.

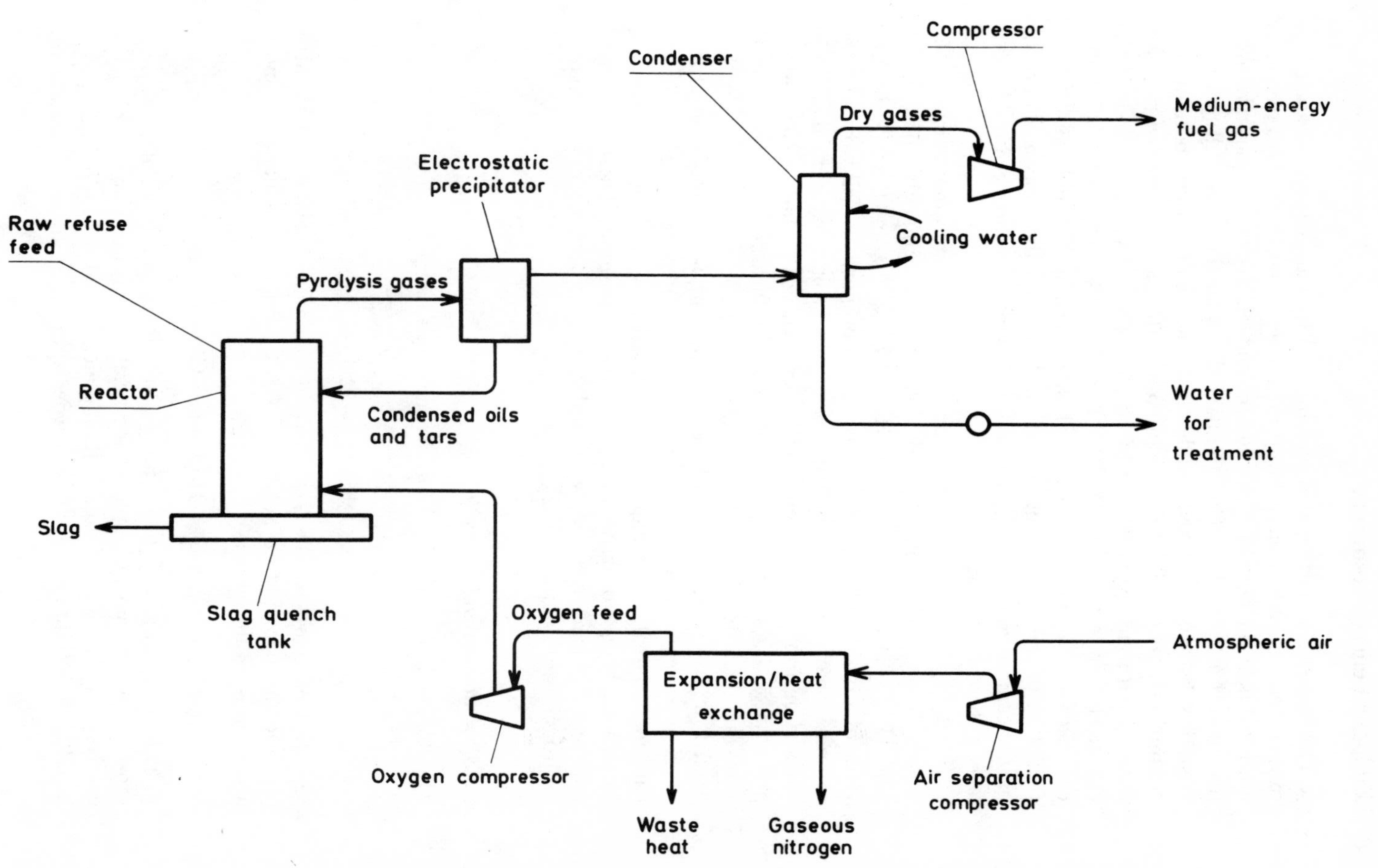

Fig. 4.14 Purox gasification process of solid organic waste (after Union Carbide Corp. [121]).

The biomass is particulated and injected into an oxygen-fed shaft, the oxygen being derived from a cryogenic air separation unit. This oxygen is admitted near the bottom of the furnace and enables the temperature to be maintained at 1600–1700°C in the partial combustion zone. All non-combustible substances are fused and descend *in toto* to the furnace bottom to be continuously removed, whilst the organic materials are degraded to gases, liquids, and solid char, the first of which rise and dry, and so preheat the charge before exiting at 205°C. The gaseous component is further enhanced by the production of carbon monoxide and hydrogen via the reduction of the low boiling liquids and char on their passage into the partial combustion zone. Higher boiling liquids are condensed on the surface of the cooler solids.

The fuel gas produced in the Purox process, at least from municipal solid wastes, consists approximately of the following percentages by volume: CO, 45.6: H_2, 27.3; CO_2, 12.8; CH_4, 8.0; H_2O, 4.7.; N_2, 1.0, and has a net heating value of around 10.5 MJ/m^3. These various proportions will vary with the composition of the feedstock. After leaving the furnace, condensed oil droplets and most of the remaining fly-ash are removed within an electrostatic precipitator, the former being recycled to the furnace for degradation to gases and the latter eliminated with the slag. The gases themselves are then freed from any hydrogen sulphide and organic acids by a neutralizing solution present within an acid absorber. Ensuing salts are fed to the furnace for removal, while moisture is removed from the saturated gas in a condenser, hence to arrive at the above composition. There is no real pollution problem envisaged, with sulphur and nitrogen oxides present only in insignificant quantities [52].

4.5.2 Substitute natural gas (SNG)

The low-medium energy gas formed may then be methanated to SNG via the reaction

$$CO + 3\,H_2 \rightarrow CH_4 + H_2O$$

Since this reaction is extremely exothermic, continuous heat removal is vital to preserve the temperature at around 400°C. Prior to methanation more hydrogen is produced via

$$CO + H_2O\ (v) \rightarrow CO_2 + H_2$$

so as to arrive at an $H_2 : CO$ ratio of approximately 3. In all of the various processes which may be employed a crucial factor is the intimate contact of catalyst with the gas stream. A Raney nickel catalyst is often used, as a metallic nickel aluminium alloy powder sprayed on to metallic surfaces, where the alloy partially melts and adheres to the surface in a hydrogen/oxygen flame. Plates or tubes may be utilized, over and through which the reaction gases circulate. Other arrangements include fixed-bed or fluidized-bed methanation reactors containing nickel or other suitable catalysts, pelleted catalyst in packed tubes, and pelleted catalyst suspended in a fluid like mineral oil. The shifting reaction of hydrogen production from carbon monoxide reacting with steam may even be combined with the methanation to SNG by using a series of adiabatic fixed-bed reactors. Whatever method is employed, any traces of sulphide are removed as far as possible, and the energy value of the product gas is raised to approximately 38 MJ/m^3. The gas stream leaves the methanator at 480°C, is cooled to just under 40°C, and finally passes to a triethylene glycol dryer where trace material reductions occur to the specifications required for pipeline transmission [52].

4.5.3 Methanol

An alternative to SNG production is the conversion of the low-medium energy gas to a liquid fuel such as methanol, thus:

$$CO + 2\,H_2 \rightarrow CH_3OH$$

This reaction is again exothermic and is encouraged by high pressure, although large-scale industrial processes have been developed operating under various conditions of pressure and type of reaction catalyst. As with SNG formation traces of sulphur are removed, and pollution problems caused by undesirable emissions are negligible. Crude methanol is condensed and separated out from the gases before purification by distillation to the final grade product requirements. The actual fuel-grade product may contain impurities such as ethanol, higher alcohols, and water, but is around 98 per cent pure [52].

4.5.4 Ammonia

Ammonia may also be formed from the original fuel gas product by firstly converting the carbon monoxide to carbon dioxide, which is eliminated, to leave an almost 100 per cent hydrogen stream. Hydrogen is also evolved simultaneously in the reaction

$$CO + H_2O \rightarrow CO_2 + H_2$$

whereby the purified gaseous mixture is reacted with an excess of steam, using an iron oxide catalyst, to bring about the diminution of carbon monoxide to less than 1 per cent. Most of this residue is removed on contact with ammoniacal cuprous formate solution. Nitrogen, obtainable from an air separation facility, is added to give a hydrogen : nitrogen ratio of 3 : 1, the gases passing to an ammonia synthesis reactor at a pressure of 30 300 kPa. The reaction temperature is 475°C, and accompanying catalysts are iron oxide with small quantities of aluminium, calcium, magnesium, and potassium oxides. The overall yield from the reaction

$$N_2 + 3\,H_2 \rightarrow 2\,NH_3$$

approaches 90 per cent of theoretical, with recirculation of unconverted nitrogen and hydrogen. Again no undesirable emissions are likely to occur to present environmental difficulties [52].

4.5.5 Electricity

The fourth and final major possibility for converting the low-medium energy fuel gases, produced from a process such as the Purox method, into a potentially more useful commodity lies in their utilization within a combined gas turbine-steam cycle to produce electricity. At a gas turbine inlet temperature of 980°C, an overall heat rate for the original biomass raw material of 12 MJ/kWh (e) would be envisaged, giving a net delivered electricity efficiency of 29.4 per cent [47], i.e. on a par with electricity derived from coal. If the biomass were to be derived from a plantation as outlined in Section 3.1, then a land area of 725 km^2 would be required to provide sufficient biomass to operate a 1 GW power station at an annual load factor of 80 per cent. This is approximately a 14.3 per cent increase in the biomass cultivation area thought necessary for direct firing as a pre-requisite to the electricity generating system (see Section 4.7). However, ammonia could conceivably be manufactured in the gasification route as a bonus by-product. Based on current technology the route via gasification is less economically attractive than that via direct combustion, although future advanced gasification and gas turbine technology could well change the picture within the next decade or two.

4.5.6 Can gasification compete?

Notwithstanding the above diversity of uses to which the gasification products could be put, each must compete with other approaches, and it would appear that probably pyrolysis (Section 4.6) and direct combustion (Section 4.7) of the biomass when dry, and anaerobic digestion (Section 4.2) when wet, are at present more efficient and do not require such intensive post-treatment procedures. Clearly, detailed energy analyses as well as economic appraisals are needed for meaningful comparisons to be made with other processes; and these should be forthcoming as the various processes concerned are further developed to the pilot-plant stage and beyond.

4.6 Pyrolysis

The second major thermochemical route to producing fuels from dry biomass is afforded by pyrolysis, whereby the organic material is destructively distilled in the absence of air (or oxygen) to yield a variety of energy-rich products. Domestic and other refuse, straw, and wood have received most attention in this area, since a moisture content of much above 35–40 per cent is detrimental to a favourable energy return. As an energy analysis of the Occidental flash pyrolysis process for converting municipal waste, primarily into oil, is included in Section 5.6, the process itself is described here with the aid of Figure 4.15. The overall scheme also employs facilities for resource recovery via magnetic separation of ferrous metals, air classification for inorganic material retrieval, froth flotation for glass recovery from the inorganics, eddy current separation of aluminium, and drying and screening of the organics to further remove the inorganics.

Before entering the pyrolysis reactor the feed, whether refuse cellulose, or indeed straw or wood, must be reduced to fine particle sizes so that high heat transfer rates may be attained for shortening the residence pyrolysis time, and thereby maximizing liquid yields. This is accomplished by shredding the material to 80 per cent smaller than 14 mesh (1200 μm) from whence it is conveyed into the reactor by recycled product gas. Flash pyrolysis occurs at 500°C and 101 kPa pressure without the presence of a catalyst. The solid char is separated from the product fluids with cyclone separators after leaving the reaction vessel. This char is then recycled to the reactor as a heat source following its combustion with air, and rapid mixing occurs within the reactor as the suspended materials pass upwards under turbulent flow. Rapid heat transfer is thus effected, minimizing further degradation of substances formed and optimizing liquid fuel yields by way of the very short residence time achieved. The vaporized material passes to a gas/oil separation unit to be quenched rapidly from 500°C to 80°C so as to prevent the large-molecule oil components from cracking further to less useful end-products. Separation is carried out into pyrolytic oil, gas, and water.

As mentioned previously, some of the outlet gases are utilized to convey shredded feed material into the reactor, the remainder being combusted in the process heater to raise the temperature of the carrier gases, char, and rotary kiln dryer. The energy content of the pyrolysis gases is around 14 MJ/m^3. A portion of the water formed is retained in the pyrolytic oil for viscosity control, with the overall products consisting of water, 10 per cent;

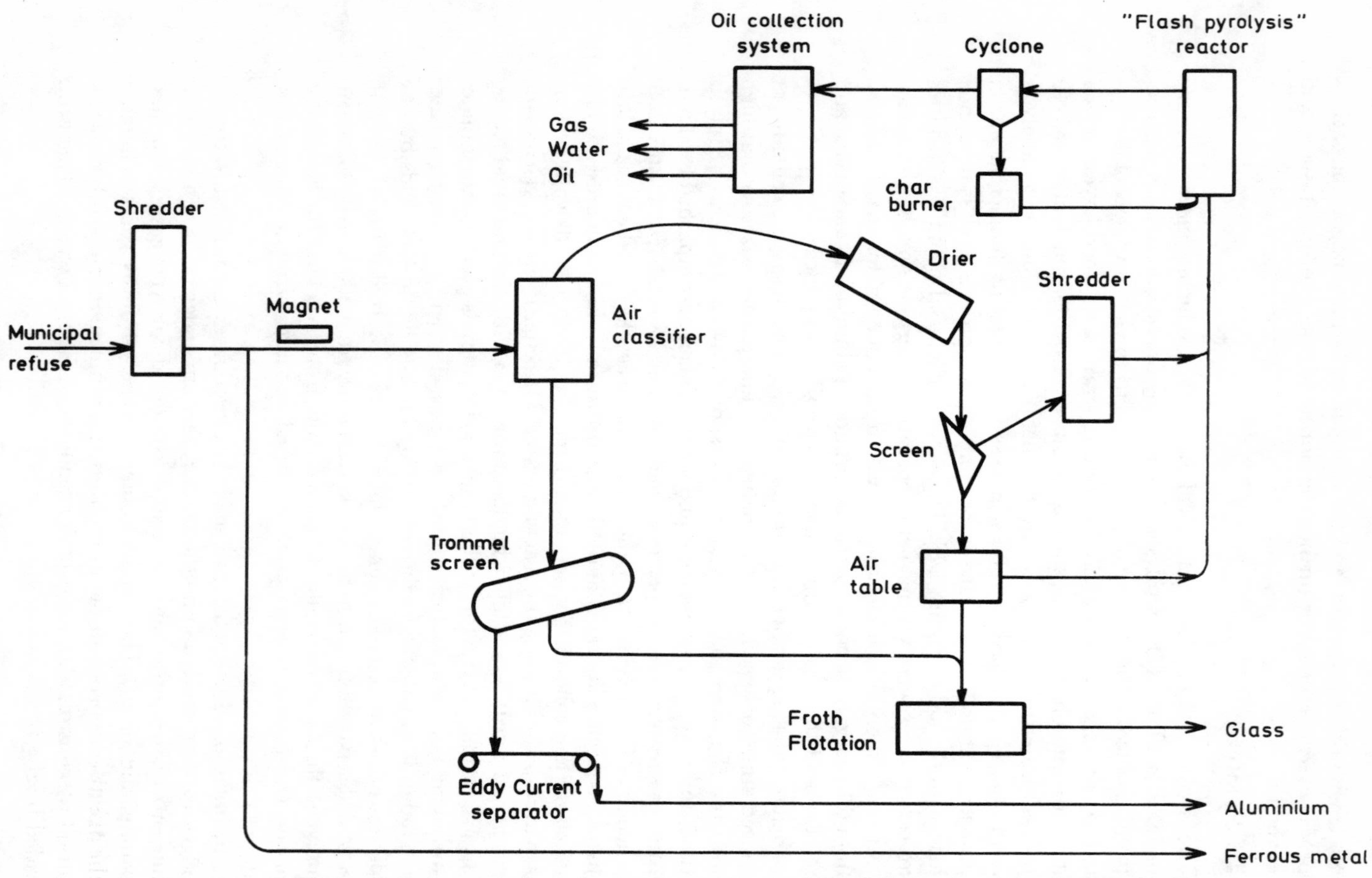

Fig. 4.15 Occidental flash pyrolysis and resource recovery of municipal refuse (after Preston [122]).

char (energy content 19 MJ/kg), 20 per cent; gas (energy content 15 MJ/m^3), 30 per cent; oil (energy content 25 MJ/kg), 40 per cent; based on the dry weight of municipal refuse feed to the pyrolysis reactor. The principal fuel formed is the oil, which is viscous, chemically complex, and has a low sulphur content. Its heating value is some 60 per cent of that of a typical No. 6 fuel oil, and it is also more viscous, requiring that it be pumped at a higher temperature, 70°C as against 45°C. Gaseous and particulate pollutants will also be emitted, the latter usually dealt with by landfilling, and the former passed through a baghouse and afterburner at 650°C for at least 0.5 sec, and then exhausted through another baghouse to the atmosphere.

The oils commonly formed via pyrolysis are invariably complex mixtures of hydrocarbons, ketones, aldehydes and other organic compounds, while the gases may contain carbon monoxide, carbon dioxide, hydrogen, ethylene, methane, ethane, and various other hydrocarbons in smaller quantities. Although the pyrolytic pressure is rarely much above atmospheric, the temperature may be at 500°C, as above, for the preferential formation of oil, or up to 700°C, when gas and char predominate. The latter is desired for the manufacture of a carbon char similar to charcoal which can be manufactured into briquettes for use as a solid fuel. However, it is probably as a conversion technology for liquid fuel formation that pyrolysis will have its major role in the future. Pyrolysis of organic materials to hydrogen and carbon monoxide would enable methanol to be synthesized in a manner akin to that described in Section 4.5, while the pyrolytic oils themselves could be purified and upgraded where required, depending upon how they were to be utilized [52, 122].

A second pyrolytic process under development, and which produces a liquid, solid, and gaseous product, is that of the 'Tech-Air Corp.', Atlanta, Georgia. This system is designed principally for the conversion of agricultural wastes such as wood chips, pine bark and sawdust, nut hulls, and cotton gin wastes, although municipal wastes and automobile wastes have also been utilized at a throughput rate thus far of 50 t per day.

A simplified flow diagram is presented as Figure 4.16. The input feed material varies in water content from as low as 20 per cent to as high as 55 per cent and the system is based on a steady-flow, low temperature processing of this material in a porous vertical bed. Included are a waste receiving unit, belt conveyer to the converter, converter and char handling system, offgas cyclone, condenser by-pass, demister, draft fan, and vortex afterburner. The dryer reduces moisture content to 4 per cent utilizing hot combustion gases from the off-gas burner. Having passed to the converter the dried material undergoes pyrolysis at a one hour residence time to yield an oil product condensed from the vapours generated, heating being provided by the admission to the vessel of process air.

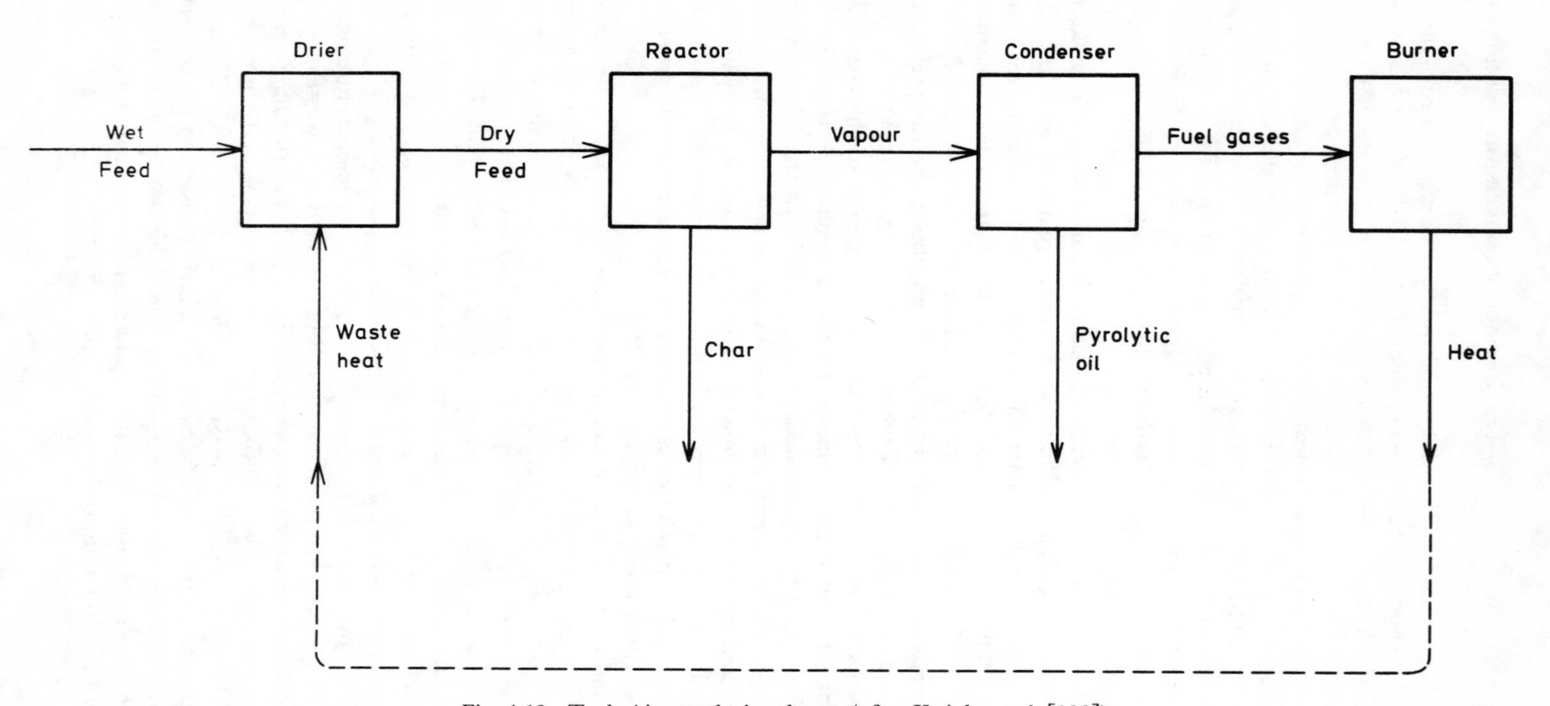

Fig. 4.16 Tech-Air pyrolysis scheme (after Knight *et al.* [123]).

The relative yields of the pyrolytic products may be varied within limits, according to input feed and operating conditions. From an essentially woody raw material 23 per cent char, 25 per cent pyrolytic oil, 68 per cent noncondensible gases, and 33 per cent water vapour are produced. The extra 49 per cent is accounted for by the nitrogen from the process air contributing to the noncondensible gases formed. On an energy basis the char and pyrolytic oil each contain 35 per cent of the dry feed energy input, the noncondensed gases, 21.7 per cent, and the remaining 8.3 per cent represents the heat loss from the system.

The heating value of the char ranges from 25.5–31.5 MJ/kg, the oil from 23–30 MJ/kg, while the noncondensed gas phase has a low energy content of 8 MJ/m^3 and should preferably be burnt on-site. It would be most advantageous to use the gas in a boiler and to recoup the waste heat from this for drying the input feed, since there is an excess of gas when the feed is heated directly [123].

Pyrolysis is thus quite a versatile process, having potential to provide a wide range of fuels from a varied number of sources with a moderate water content. Experience of large-scale operations is increasingly being obtained, efficiencies are improving, and there is little problem with polluting by-products. Perhaps the two major constraints are that the products of pyrolysis are not always entirely predictable, and that the reasonably high temperatures involved combined with the possible formation of corrosive compounds may lead to a shortening of the reaction vessel's life-time. The first of these is the more important since, for the most part, the necessary equipment is not elaborate and can be replaced with relatively small expense.

4.7 Direct combustion

The final major conversion process is the familiar one of direct combustion where heat is the basic product, produced as high-pressure and high-temperature steam derived from the biomass' heat of combustion. Pre-drying, to varying degrees, is normally required to achieve the necessary low water content of not much more than 15 per cent, and even the most commonly burnt material, wood, is no exception here.

At the outset it can be argued that drying, other than by the sun, together with present process inefficiencies and the probably more profitable products formed by other conversion methods, conspire to render the manufacture of heat as an irrelevant luxury. However, this will be dictated by circumstances, and in the case of recuperative incineration of municipal refuse, where the addition of a heat recovery boiler connected to a refuse incinerating plant is possible, energy may be reclaimed which would otherwise be wasted. The heat recovered, at around 55 per cent boiler efficiencies, provides a net energy output of 5.5–6.5 MJ/kg of refuse input, and may be utilized in district heating schemes for the heating of industrial, domestic, commercial, and institutional buildings. Steam raising for heat distribution by such methods may well become more economical if energy prices rise in real terms, and district heating is already practised in many countries of the world, particularly so in Scandinavia, and especially in Sweden [124]. However, it is perfectly true that these organic wastes might well be preferentially treated by pyrolysis or anaerobic digestion to yield fuels which are more easily stored and transported, and so such raw materials as wood and bagasse will be concentrated upon in this section.

As well as its use as process or saleable heat any generated steam can, of course, be used for the generation of electricity. Beginning with steam generation for on-site and industrial use, fluid-bed systems, akin to the proposed fossil fuel systems utilizing coal and natural gas, are gaining popularity, with wood as the combusted material input. Here, fuel combustion results in the evolution of high-temperature gas which is transported to a boiler, surrenders its heat for steam raising, and is finally discharged to the stack. In this way the separate functions of combustion of the biomass and the subsequent generation of steam are kept apart for the easier adaptability of the various possible inputs.

Low-grade wood wastes of a range of moisture contents, sizes, and containing non-combustible matter, as well as carbon-containing fly-ash from more conventional wood boilers, are currently the most common inputs to fluidized bed systems, the sizes of which are relative to the raw material availability. The fluid bed itself is prepared with sand or similar substances, and the raw material waste is admitted gradually at a uniform rate, with the combustion air acting as the fluidizing medium. The bed temperature is kept usually between 650°C and 1000°C, well beneath the fusion or melting point of the bed material, by regulating the excess air and moisture content. System efficiencies are improved by immersing heat absorbing tube banks within the bed to allow a reduction in excess air, while at the same time not increasing bed temperatures. Provided capacity for steam generation above 275 t/hour can be accommodated, that is, large volumes of steam, the technique could supersede electricity production from the conventional stoker-fired boilers described below [52].

Whereas thermal energy produced in fluidized beds is subsequently recovered primarily by convection, that from more conventional boiler-furnace generation is removed via radiant transfer to a heat receptive surface and then by convection heat transfer. A familiar type of steam generator would be of the water-tube variety for dealing with bagasse, wood, and/or bark, in which the fuel is spread across the combustion chamber on to a travelling grate. A travelling grate effects better cleaning of the resulting ash than other grate systems by continuous dumping, thus ensuring a longer grate life. The feeding operation, whereby small particles burn in suspension while larger ones drop on to the grate, spreads a thin, even layer of fuel, and over the grates a flame radiates heat back to the fuel as an aid to the combustion process. The character of combustion is controlled via underfired and overfired air, and the unit may possess a tubular air heater for preheating the combustion air. As there is little refractory material present, the furnace, whose walls are lined with heat exchange tubes, can adjust rapidly to variations in load. The spreader stoker firing method adopted here is the most prevalent type for wood-firing boilers in the USA, while the most common output capacity lies in the range of from 7 t to 45 t steam per hour. The maximum size of such a steam generator is in fact 275 t/hour, but this capacity could be doubled using a double grate incorporated side-by-side. Steam can be generated up to a temperature of 480°C and a pressure of $9.3 \times 10^6 Nm^{-2}$, with the energy liberated from 5–8 cm sized wood at 55 per cent original water content as much as 110 MJ per m^2 of grate area. With respect to polluting emissions, sulphur dioxide is not produced in any great amounts, although carbon monoxide, oxides of nitrogen, and unburnt hydrocarbon gases could present problems [52].

Electricity generation from wood or bagasse is carried out locally in the respective timber and sugar industries, but for large-scale central power station electricity utilizing a woody feed several developmental issues need to be resolved. To draw an analogy with a 1 GW coal-fired station where the coal requires to be greatly pulverized, only sawdust and sanding dust are of a comparable size and are costly to manufacture. The energy consumption involved for such massive wood pulverization might indeed make the whole concept unworkable, and similarly the heat consumed in wood drying, and indeed the transportation energy over moderate to long distances, all combine against the utilization of wood fuel, notwithstanding its low sulphur and ash content, as a producer of electricity on a large scale. Nevertheless, if such a scheme did arise its basic features would differ little from the coal-fired steam turbine electricity generator layout as shown in Figure 4.17. The steam produced as outlined above is done so at as high a temperature as possible to improve the overall thermodynamic efficiency; it drives the turbine, and continues to the condenser where it is condensed to water before being returned to the boiler. In turn the turbine drives an alternator (AC generator), the output from which is transformed to a high voltage, and transmitted to the final user, where the voltage is decreased to the necessary value concomitant to its convenient utilization [52].

Section 4.5 contained an appraisal of the possibilities for electric power

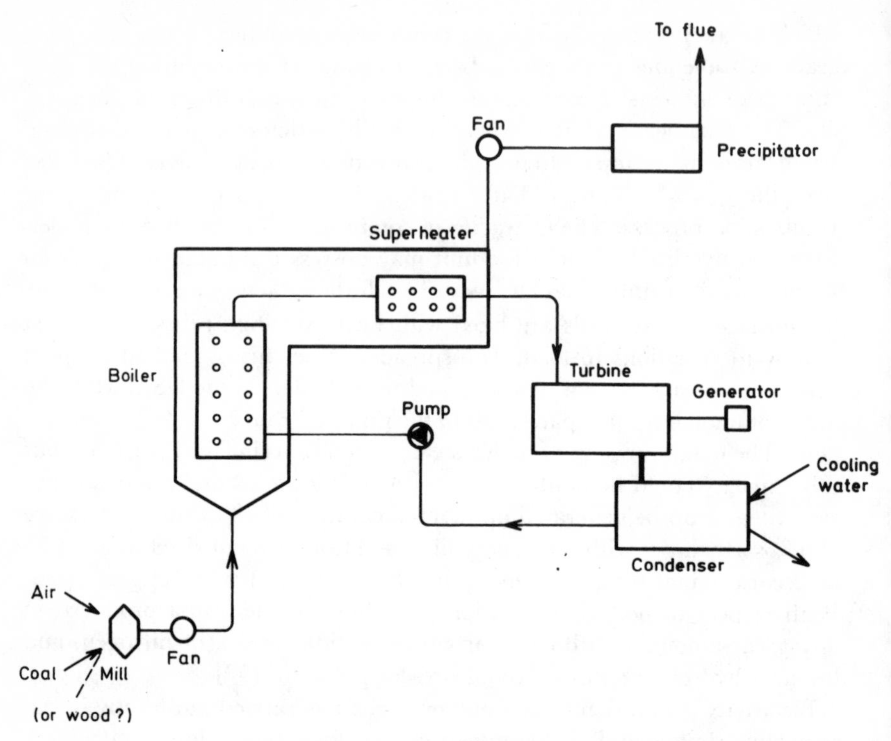

Fig. 4.17 Steam turbine electricity generation.

production via the gasification of biomass to a low-energy gas, where the biomass had been cultivated within an energy plantation [47]. The advantage of direct combustion here would be, for a 1 GW power station at an annual load factor of 80 per cent, an appreciable reduction in land area requirement for biomass growth. Thus, if a large-scale system on these lines were brought into operation then direct combustion would appear to be the method most favourable on an output : input basis. The economic costs of generating electricity by the direct combustion of wood are variable, depending naturally on specific local situations, but are thought to be competitive even today with oil- or gas-fired plants in some areas of the United States. Chapter 6 will cover this and other fuels-from-biomass schemes from the economic viewpoint in greater detail.

Chapter 5

The energetics of biomass systems

5.1 Net energy

One need hardly labour the point that there is no useful purpose in developing an energy source whose production consumes as much energy as it produces. Yet, there have been many proposals for bioenergy sources in which that is the outcome. Let us take a particularly common example – the 'biogas' generator in a temperate climate. The concept is simple. Human and animal faeces are anaerobically digested at about 35°C, yielding a combustible gas–methane mixed with carbon dioxide. This process does not work unless two conditions are satisfied: the mixture must not fall below 30°C and it must be stirred. In a tropical environment the first condition is easily met; in temperate climates, unless the scale of operation is very large (that is to say the digester has a large volume relative to its heat losing surface area) heat must be applied. The obvious fuel is the digester gas itself. It is all too easy to choose a scale of operation where heating requirements alone can consume the entire output of gas. But even when this is not the case, as in large digesters, the energy for mixing and stirring usually effected by an electrical motor, can consume so much power, as to once again require more energy than it puts out.

It is therefore a useful and necessary exercise to find out the net energy of an energy producing system. Such calculations are, however, complicated by the fact that not all energy sources are equally effective. For example a kilogram of fuel oil will, on combustion with a stoichiometric quantity of air, yield 45 MJ of heat. The same amount of heat may be obtained from 1.5 kg of hard coal and 2.5 kg of lignite, and about 3 kg of wood. Yet the heat from the 1 kg of oil, though the same 45 MJ as from 3 kg of wood, is much more effective, because it is at a higher temperature. Thermodynamically speaking 45 MJ of heat from the combustion of oil can do much more work than 45 MJ from the combustion of wood.

Thus, returning to the example of the 'biogas' generator, 45 MJ of 'biogas' heat are worth less, as work, than even 45 MJ of wood heat.*

Consider the following hypothetical example of a 'biogas' generator producing 100 MJ of 'biogas' heat per day. Let us assume it is well insulated so that no heat is required, but it has to be stirred with a 0.5 hp electric motor. What is the net energy of the system? There are two ways one can make this calculation. The rigorous method is to express the outputs and inputs to the system in terms not of heat, but that which engineering thermodynamicists call 'available work'. Such an approach calls for a degree of expertise in thermodynamics that escapes most workers in this field. Let us therefore pursue a less rigorous, but nonetheless accurate method.

The first step is to decide the life-time of the plant, and then sum all the inputs and outputs over that life-time. Thus, the inputs will be:

material of construction
energy of construction
maintenance materials
electricity to the stirrer

The outputs will be:

'biogas' – 100 MJ/day
soil-conditioner

Our first task is to assess the plant life-time. Let us say 30 years. The second task is to assess the maintenance requirements. Let us assess these at 100 per cent of the original construction costs, thus effectively reducing life-time to 15 years.

To establish the net energy of the 'biogas' generator system, we now need to know how much energy was required to make the inputs, moreover, not just energy, but 'biogas' energy. Now the steel and concrete to make the 'biogas' generator were not made with 'biogas', but coal, oil and other higher grade energies. However, since these commodities are made largely with heat, and use comparatively little work, we can content ourselves with ascertaining the heat requirements of making them. For the moment let us ignore how this number is obtained and simply take the authors' word for it that an appropriate figure would be 15 000 MJ. The electricity to the 0.5 hp motor would be, allowing for efficiency losses, about 0.5 kWh/hour, whose production, using 'biogas', would consume about 7.2 MJ of 'biogas' heat.

*Though the end-use of gas is more efficient than that of wood.

However, operation is intermittent, let us say 50 per cent of the total time.

Thus we have a life-time balance-sheet for 15 years of:

Energy in:
construction 15 000 MJ
stirrer:

$$15 \text{ years} \times 8760 \frac{\text{hours}}{\text{year}} \times \frac{7.2 \text{ MJ}}{\text{hour}} \times \frac{1}{2} = \frac{473\,040 \text{ MJ}}{488\,040 \text{ MJ}} (\text{total})$$

Energy out:

$$15 \text{ years} \times 365 \frac{\text{days}}{\text{year}} \times \frac{100 \text{ MJ}}{\text{day}} = 547\,500 \text{ MJ} (\text{total})$$

$$\therefore \text{Net energy over life-time} = 59.460 \text{ MJ}$$

or 10.9 MJ/day:

equivalent to about 0.75 kg of air-dried fuelwood [125].

Such a calculation gives one leave to doubt whether the complication of a 'biogas' generator is worth the mere 10.9 MJ of net heat generated each day, for no account has been taken of labour, either for construction or subsequent management. But of course other arguments may prevail. Firstly, the system is a positive, albeit small, net energy generator. Secondly it disposes of a polluting material–faeces. Thirdly it produces a soil-conditioner. If this latter is of value, then under the rules of net energy analysis, we should credit to the 'biogas' generator that amount of energy (as 'biogas') that would have provided the same service (that is to say, the same soil-conditioning effect). Let us suppose the annual output is equivalent to 100 kg of nitrogen, in the form of ammonia. This would, if made and transported to the site have required an energy (as 'biogas' heat) of about 60 MJ/kg. Thus to the output we should credit $100 \times 60 = 6000$ MJ/year or 90 000 over the life-time. This substantially improves the picture to a net output of 149 460 MJ or 27.3 MJ/day. Finally, if a boy or boys can be employed to do the stirring, then the net energy is hugely increased to the equivalent of about 7 kg wood per day, but of course, the boys have to be fed!

The reader anxious to carry out net energy analysis is advised to consult an appropriate text, such as *Energy in the Economy* by Slesser [126] or the short report summarizing the conventions of energy analysis published by the International Federation of Institutes for Advanced Study (IFIAS) in Stockholm [127].

Throughout this text we shall use the IFIAS convention. In this the unit of computation is the Joule (usually as Mega Joule: MJ = 10^6 J) expressed as the Gross Energy Requirement (GER) to deliver a good or service to a particular point of the economic system. GER implies the heat equivalent of the primary energy resources (non-renewable) that is needed to furnish the fuels which create the aforementioned good or service.

Another way of examining net energy is by the NUEP or Net Utilizable Energy Production [128–130]. It is defined as

$$\frac{\text{externally utilizable energy} - \text{secondary energy input}}{\text{primary energy available (raw materials)}} \times 100$$

This method counts both renewable biomass and non-renewable energy sources within the same parameter, and so similar NUEP numbers can have quite different underlying implications. For example, let us take the information in Table 5.7 for ethanol derived from sugar cane, the sugar cane itself being produced at an intensity of 70 tonnes/ha yr. The externally utilizable energy is 1 kg of ethanol, having a gross calorific value of 29.7 MJ.

The secondary input, that is to say the gross energy requirement of the system, is 24 MJ.

The primary energy available, which in this case is essentially sugar, but should include some bagasse, can be taken at 16.4 MJ per kg of sugar, or 36 MJ for 2.2 kg of sugar required.

$$\text{Thus, NUEP} = \frac{29.7 - 24}{36} = 16 \text{ per cent.}$$

Now suppose that we reduced the gross energy requirement for the process, because we decided to utilize sugar as a fuel in place of the fuel oil. To replace the 8 MJ of fuel oil required, per kg of ethanol produced, 9 MJ of equivalent sugar energy would be needed, allowing for a lower quality of sugar energy compared to fuel oil energy. However, to make this 9 MJ of sugar energy would require 0.55 kg of sugar, making the demand for sugar for the system 2.75 kg per kg ethanol produced. The NUEP will therefore be

$$\frac{29.7 - 17.2}{2.75 \times 16.4} = 28 \text{ per cent}$$

from which we should conclude that the system is more efficient. However, if we now work it out on the basis of net energy per hectare, we get in the first place 97 GJ/ha yr as the net energy output, whereas in the second case we get 170 GJ/ha yr net energy. We see here then an increase in the net energy per hectare, perhaps a surprising result in view of the decrease in productivity per hectare. Let us take the data one step further and assume that we generate the electricity by the combustion of sugar in an electricity generating station. The sugar itself must be manufactured from a system whose intensity is a GER of 2.2 MJ per kg of sugar. Under these circumstances we find that the net energy has now risen to 218 GJ/ha yr, while the NUEP efficiency has become 36 per cent.

Both forms clearly have their value, but the absolute number as given in the net energy per hectare-year is, we believe, a more useful way of conveying the economic value of the system under study.

5.2 The energy crop plantation

Section 5.1 demonstrated the importance of establishing the net energy of a biomass energy system. Yet, net energy is not the only criterion of interest. While, clearly, there is no point in creating a biomass energy system which consumes as much energy as it produces, the mere fact that it is a net energy producer is not of itself a sufficient reason for using such a system. The reason for this is that biomass systems represent a roundabout way of utilizing solar energy, and therefore they are land-dependant. We can identify therefore two important criteria in assessing biomass energy systems: their net energy yield and the land requirement for the system. Table 5.1 indicates a number of systems that either exist in reality or have been postulated. We can see that their net energy varies enormously, and that their net energy per hectare of land utilized puts them in different ranking order. Clearly, the important criterion is to obtain the maximum net energy per unit of land area utilized. However, having said this, one must bear in mind that just as there are some energies which are more valuable than others on a Joule per Joule basis, some areas of land are similarly more valuable than others. The relation between net energy/hectare and cost is dealt with in Section 6.2.

We get some feeling for this problem by considering the net energy of food production in the UK. It has been calculated that on average it takes about 1.75 units of energy, expressed as Joules of heat, to create at the farm gate one unit of food energy and that the trend of increasing energy intensity within UK agriculture continues [133–135]. If that were looked at in the context of a biomass energy system, then such a system is a non-starter. It uses more energy for its production than it will produce in its consumption.

However, we do not produce foods, whether for animals or people, in order to evaluate them by their heat of combustion, but rather by what they can do for us both in terms of palate and nutrition. The fact, therefore, that the net energy of the UK's food producing system is negative, is not of itself a serious or even important piece of information. On the other hand, if we

Table 5.1 Biomass production and its conversion to fuels (adapted from [131])

Principal substrate	Product	Productivity (t/ha yr)	Productivity (GJ/ha yr)	Net energy (GJ/ha yr)	Net energy (GJ/t)	GER product (GJ/t)	Man-hours/t	Key inputs
Atmospheric CO_2	energy crops	68	1175	+1090	+ 16	1.26	0.3	water[a] fertilizer, fossil fuel
Raw sewage	algae[b]	25	575	— 850	— 34	57	34	fossil fuel
Raw sewage	algae[c]	25	575	+ 125	+ 5	18	34	fossil fuel, flocculants
Algae	methane[c]	5.6	269	— 672	—112	168	200	fossil fuel
Livestock waste (UK)	methane	0.01	0.56	— 0.88	— 88	144	237	fossil fuel
Sugar cane*	ethanol	17	505	+ 97	+ 5.7	24	6.8	N,P, fossil fuel
Cassava*	ethanol	2.1	63	— 65	— 31	61	25	N,P. fossil fuel
Timber	ethanol[d]	0.35[f]	10.5	— 73	—209	239	21	N,P. fossil fuel
Timber	ethanol[e]	0.22[f]	6.6	— 15	— 68	98	120	N,P.H_2SO_4, fossil fuel
Straw	ethanol	0.7	20.8	— 136	—191	222	20	N,P. fossil fuel

[a]Water availability is assumed to be sufficient
[b]The figures relate to current methods adopted
[c]The figures are estimates of what should be possible within a few years or even at present
[d]Fermentable sugars formed by enzymatic hydrolysis
[e]Fermentable sugars formed by acid hydrolysis
[f]Productivity expressed on basis of annually renewable wood quantity per unit area for use as substrate
*The most recent Brazilian energy analyses indicate high positive net energy returns via the use of either hydroelectricity and/or wood and bagasse as the principal fuel inputs [20, 21, 132].

would look at biomass systems and find that they had a zero or negative net energy, then we can instantly reject them, even if they are being created on marginal or low quality land. They are, effectively, drawing more energy from the economic system than they are replacing. On the other hand, if we discover a biomass product that has a positive net energy, it is of interest to us, but we must calibrate it against the land area utilized or sequestered by that biomass system. Let us suppose, for example, that a plantation of hazel wood is operated as a coppice energy plantation. Let us suppose that it grows in marginal land, does not use good or even moderate agricultural land, but land which would otherwise lie fallow and that it yields without any inputs whatsoever the equivalent of one tonne of dry wood per year. We will, for the moment, ignore the energy requirements of cutting, assuming that this is done by a boy using a handsaw. One tonne of this dry wood represents about 15 GJ of heat. Supposing now we intensify the growth of that coppice by the application of fertilizer, then we are in the business of making a mass balance and an energy balance on the same lines as that used for the 'biogas' generator in Section 5.1. We would find that modest applications of fertilizer, even allowing for the energy requirement for creating and distributing that fertilizer, would yield a larger net energy from the whole coppice system. If, then, we continued to intensify the application of fertilizer, the system would peak out and thereafter there would be an erosion of the net energy product from the system. If we would retransfer the coppice from the marginal land, to let us say, good argicultural land, we would get a different yield profile. The point of maximum net energy return would undoubtedly be at a different level of fertilizer intensity. In judging the most economic level of fertilizer application, we would find that the economic decision rested on such factors as land value as well as net energy, and net energy per unit land area. Therefore, a primary calculation to be made is to establish the net energy yield per hectare as a function of intensification of inputs, and set this against the relative value of the land visualized for the process.

This concept has important ramifications, because if we choose to express net energy as a ratio, as many researchers do, then without doubt one can obtain the highest ratios by putting in zero inputs, and simply utilizing solar energy. A natural coppice without any fertilizer addition, with no energy-driven machinery gives you an output of 15 GJ for an input of zero. As a ratio that is infinity. If, by the application of fertilizer whose energy content is equivalent to 5 GJ, we raise the output of the coppice to 25 GJ, then the net output rises from 15 GJ/ha to 20, but the energy ratio falls from infinity to 25/5, namely 5. Further addition of 5 GJ of fertilizer input might raise the system output to 32 GJ. The net energy of the system is now (32–10 GJ), which is 22 GJ/ha per year. This is still an improvement on the previous level of intensification which gave a net energy of 20 GJ/ha yr. However, now the net energy ratio has fallen from its original value of infinity to 32/10, namely 3.2. We see clearly therefore that net energy ratio is not a number of real interest. The number of interest is the net energy of the system per hectare per unit time, and subsequent economic calculations must take that factor into account as well as a value being placed on the land in use.

Having now dealt with an abstract example, let us turn to a proposed energy plantation from a very large scale and see what its original protagonists visualized as the system output. We take the case of a hypothetical energy crop plantation postulated by two Stanford Research Institute workers, John Alich Jnr and Robert Inman for a 272 000 tonnes/year dry weight biomass system using a polyculture of short season crops and perennial plants harvested several times per growing season. Their balance sheet is summarized in Table 5.2. We have taken their data so far as material inputs are concerned and have used our data on energy equivalence to establish the net energy yield. We see that the energy ratio works out as 14 units of energy yielded per unit of energy put in (this should be compared with the original 18–1 calculated by the SRI researchers using a different energy analysis methodology). The net energy gain works out at 15.98 MJ for each kilogram of biomass produced; that is to say 1 kg of biomass has an energy equivalent of 17.24 MJ, while it took 1.26 MJ to create that kilogram. However, by far the most important number is the fact that the yield postulated is no less than 1087 GJ/ha yr, a much higher figure than is obtainable by any other proposed system. But there is a foot-note (a) in Table 5.2 which has a hidden significance, because the crops proposed will require a great deal of water for their growth. The Stanford Research Institute solution was to grow the crops in an arid or sunny area like New Mexico or south-central Texas, but where the water requirement of the crops would be greatly in excess of the natural rainfall of the region. If we visualized, for example, that the energy plantation was located in south-central Texas, and that the water balance was maintained by pumping water from the Mississippi River at some 500–700 km distance, we would have to ascribe to the energy requirements for the crop around 7.7×10^5 GJ annually without even considering the capital expenditures to bring that quantity of water to the biomass plantation. This increases the calculated energy requirement of growing the biomass from 1.26 to 4.09 MJ per kg and reduces the net energy from 16 to 13 MJ per kg, equivalent to a net energy yield of around 880 GJ/ha yr. While this is still a very large figure compared to any existing known system it represents a significant reduction on the figure of 1087 GJ/ha yr.

Table 5.2 Inputs per 272 000 tonnes dry biomass/annum (land area 4000 ha; labour force 44)

Major physical inputs	*kg/t Dry biomass*	*GER: MJ/t Dry biomass*
NH_3	16.68	844.0
P_2O_5	0.84	5.9
K_2O	1.68	16.1
Herbicides	0.05	5.3
Insecticides	0.025	3.3
Fungicides	0.017	1.8
Water	227×10^3	—(a)
Fuel oil (for herbicide, pesticide, and fertilizer application, planting, cutting, chopping, fresh hauling, turning, drying, dry hauling)	3.36	155.6
Steel (b)	0.24	21.8
Electricity (98% for irrigation)	14.34 kWh(e)/t biomass	200.8
Seed production (0.3% of above total)		3.8

GER dry plant biomass = 1258 MJ/t = 1.26 MJ/kg; gross energy content of biomass = 17.24 MJ/kg; net energy gain in biomass = 15.98 MJ/kg or 1087 GJ/ha yr. Energy input : energy output = approximately 1 : 14 (compared to the SRI estimate of 1 : 18).

a The water requirement is put at 1500 mm/yr rainfall, equivalent, and this is assumed to accrue from natural rainfall, with no GER assigned to it.

b The steel is classed as a capital input with amortization rate from 2–10 yrs, depending on the type of farm machinery of which it is a constituent (e.g. tractors, planters, fertilizers, herbicide turners, haulers, pumps, combine harvesters).

We now proceed to examine a number of proposed systems in the light of this method of analysis without, at this moment, considering their economics.

An energy analysis of the silvicultural biomass farms at two US sites giving net energy outputs of 143 and 445 GJ/ha yr, also described in Section 3.1, are presented below (Table 5.3) as due to Robert Inman [48]. He assumes the same level of annual production in both cases. Thus, the energy gained through the biomass farming is greater in Louisiana than in Wisconsin, reflecting differences in water needs, land quality and climate. Unlike the example in Table 5.2 no provision has been made for the capital inputs required in the manufacture of the farming equipment and its subsequent amortization. This energy expenditure on such equipment

Table 5.3 Energy consumption and balances for silviculture plantation fields at 2 US sites – Wisconsin and Louisiana

		Energy consumed (TJ/yr)	
Operation	*Material*	*Wisconsin*	*Louisiana*
Supervision	gasoline	1.76	1.76
Field supply	diesel/gasoline	0.56	0.56
Harvesters	diesel	9.95	8.30
Tractor haul	diesel	5.75	5.79
Loading	diesel	3.50	3.47
Transportation	diesel	30.42	18.27
Irrigation move	diesel	1.37	0.69
Irrigation pumping	diesel	235.95	117.99
Manufacture	urea	104.83	107.96
Manufacture	P_2O_5	8.47	8.62
Manufacture	K_2O	16.27	18.02
Ground operations	diesel	1.22	0.06
Aircraft operations	gasoline	0.16	0.17
Fertilizer transport	diesel (rail)	1.03	0.68
Total energy consumption		421	292
Total energy yield		4484	4484
Area site : ha		28 400	9415
Net energy yield : GJ/ha yr		143	445

If the capital energy inputs for farm machinery are included at the level stated previously, then the energy consumed for the Wisconsin and Louisiana operations rises to 428 TJ/year, and 297 TJ/year, respectively.

was 1.73 per cent of the final GER in Table 5.2. If a similar proportion were assigned in Table 5.3 it would make a negligible contribution to the overall energy balance. Irrigation, once more is seen to be the highest operational energy-consumer, followed by the provision of nitrogen fertilizer. With future technological improvements envisaged these calculated net energy yields could be even more favourable in the years to come.

5.3 Low intensity biomass systems

Both Tables 5.2 and 5.3 present input details of intensive biomass cultivation systems, where significant quantities of energy are expended in order to provide high yields per unit land area. This is the typical approach of a modern industrialized society. But, it is questionable whether this is necessarily an intelligent approach to the problem. May it not be a misplaced allocation of resources, which in an industrialized society might be better directed towards more technologically oriented renewable energy resources, such as photovoltaic electricity? At the other end of the spectrum are the natural ecosystems, in which the only impetus for plant growth is solar energy and those nutrients plus water found on site. Table 5.4 below indicates some of the yields obtainable from such systems. Where the population density is low, the provision of energy from such natural systems may, indeed, be perfectly acceptable, or even desirable. Yet, all over the world, there is a tendency for populations to grow, and therefore population density to increase. This, in turn, puts a pressure on the natural biomass production. We see it all too clearly in the fact that today in many Third World countries there is an impending energy crisis in terms of fuel wood, and that the natural growth of forests and coppices no longer meets the demands made upon these natural sources. The question development planners face in the Third World countries is whether to seek a solution along the intensification of the natural solar energy driven ecosystems, or whether to abandon this method of energy provision, and go directly to the methods adopted by the industrialized countries, namely the extraction of oil or the provision of intensive energy through, for example, nuclear reactors. The problem is made more difficult by the fact that the first option, namely the intensification of natural ecosystems calls for both capital inputs and a better management. Very often, in Third World situations, better

Table 5.4 Output of selected natural and other systems (adapted from [135]).

	Energy ratio (output/fossil fuel input)	*Net energy GJ/ha yr*	*Output GJ/ha yr*
Kung Bushman, Kalahari Desert	infinite	0.003	0.003
Shifting cultivators, Congo*	infinite	5.2	5.2
Tsenbaga tribe, New Guinea – yams*	infinite	0.47	0.47
– sweet potatoes	infinite	0.40	0.4
Dodo tribes, Uganda	5.0	0.64	0.8
Subsistence agriculture, India	14.8	9.6	10.3
Rice, Tanzania	23.4	3.6	3.8
Corn, Mexico	30.6	28.5	29.4
Peasant farming, China	41.1	274	281.0

*Utilizes fertilization obtained by burning down forest and then moving on after 3–4 years.

management is hard to acquire either through lack of social organization or lack of technological training. The second approach, that of the industrialized countries, is in many ways extremely attractive, requiring very little capital input, requiring very little management and acquiring utilization of simple devices of limited life like, let us say, an irrigation pump. However, communities developed through intensive energy sources like oil or nuclear energy, are clearly more sensitive to changes in energy prices than those which build up their own intensified biomass systems. In our examination of the problem then, it will be useful to see how the transition from a natural ecosystem providing biomass to a more intensive biomass system can be achieved and what are the resulting benefits.

Table 5.4 depicts some nine energy systems of varying degrees of intensification. In the first place, the productivity is examined in terms of their metabolizable food energy product rather than their biomass energy content, for in this way we can see more clearly the impact of intensification.

The range is clearly quite enormous. On the arid Kalahari desert a Kung bushman finds his food where it grows, making substantial journeys in search for it. The energy ratio of such a system is infinite, for the inputs other than solar energy are zero. However, such a system has a trivial output, something like 0.003 GJ of metabolizable energy for every hectare each year. Better values are in fact obtained by tribes living in jungle areas who fell some of their forest woodland on a rotational basis to grow crops, as is the case with the Tsenbaga tribe in New Guinea. The felled trees are allowed to dry and are then burned giving a highly fertilized soil, which is slowly depleted over some subsequent years and finally abandoned. Since the fertilizer comes from the combustion of forest, and the forest is itself a renewable energy resource using only energy from the sun, the non-renewable inputs to the system remain zero. However, by better management and by living in a more productive area, namely where water is not in short supply as in the Kalahari desert, the output rises by a factor of over 100 to around 0.4 GJ/ha yr. Better management and the use of animals and the recycling of animal and human wastes can even further intensify the output, and figures as high as 5 GJ/ha yr have been computed for shifting cultivators in the Congo.

Table 5.4 then goes on to list a number of marginal agricultural systems in which some non-renewable energy resources are utilized to stimulate the system. We see enormous increases in the metabolizable energy output per hectare, but of course a corresponding reduction in the energy ratio. However, as we have remarked earlier, the energy ratio is not really an interesting number. What is interesting is column 2, which depicts the net energy of the system in GJ/ha yr. The clear supremacy of well-managed intensively cultivated peasant farming may be seen. However, of course this is not the entire story, because we do not know from these figures what is the populaton density in each case. If, for example, each member of a Tsenbaga tribe has at his disposal five hectares of land, then he has effectively 2.3 GJ of metabolizable energy available each year, enough for survival if not to grow fat on. By contrast, if in the Chinese situation there were 0.3 hectares of agricultural land *per capita*, a net energy of 274 GJ/ha yr would produce a fantastic output of 82 GJ per year per person. This would allow ample for conversion into animal protein and would provide the Chinese with a very high-grade diet. In fact, of course, only a small proportion of the Chinese agricultural communes reach this level of production and intensification.

The figures in Table 5.4 should now be re-worked in terms of biomass production.

The paucity of information on low intensity biomass systems makes it difficult to provide a framework of analysis at this moment. The only advice that may be given is that those whose business it is to consider whether a Third World rural type development should take place along biomass lines, or along the importation of energy intensive energy sources should obtain real-life data from the areas under consideration. The analyses should be made upon a rigorous energy analysis procedure, such as has already been indicated, and every care taken to ensure that all types of inputs are taken into account, and that the numbers are expressed in terms of the quality of of the output energy.

Ideally, a study would be pursued in the following way. The region to be examined would, we presume, have a number of natural ecosystems whose productivity should be assessed. Then a number of experiments must be embarked upon in which the natural systems are intensified using the best possible advice. For example, there is no point in intensifying a forest system purely by the application of nitrogen fertilizer and omitting phosphorous. It is then necessary to implement a programme to examine the various routes to intensification. Here, one may justifiably look at those already obtained in more developed countries; for example, the plans of the Brazilians to grow sugar cane intensively as a fuel alcohol source. However, today there is not an adequate information base upon which decisions concerning energy systems for Third World countries can accurately be made. We now turn to a consideration of those few for which there does exist some sound data base.

5.4 Methane via anaerobic digestion

Methane production via anaerobic digestion of sewage-grown algae, livestock waste, and municipal solid waste has been described in previous sections. Although energy analyses have been performed on such systems they rarely give consistent results, particularly in the second case, because of differing local conditions with respect to climate and level of technology employed. Even the values for the Californian algae-methane systems will change with time as the process is further optimized.

At one time methane production through algal growth on sewage was claimed to be a potentially useful energy source. Let us consider the published facts. The GER of algal growth [131] using recent methods of cultivation was around 57 MJ/kg (dry weight), the separation process by centrifugation alone contributing an energy input of 42 MJ/kg, equivalent to 74 per cent of the total GER. Included in the original figure of 57 MJ/kg was an energy credit of 4.3 MJ/kg, allowed since the sewage would have incurred a GER for its treatment in any event; but even so, taking a mean algal volatile matter energy content of 23.2 MJ/kg, the growth and harvesting operations alone were a net energy loss. Of course if the algae were considered for use primarily as a protein-rich animal feedstuff, then the high GER is acceptable. However, within the context of the algae as a methanogenic substrate a net energy surplus is essential, and hence the GER of the process must be lowered in order for the total process to be at all feasible. Thus, the species control and subsequent low-energy intensive harvesting procedure outlined in Section 3.1 can go a long way to reducing the GER, perhaps by 25 MJ/kg dried algae. In addition, the drying may be dispensed with if the cells can undergo anaerobic digestion in the wet form. This could result in a further substantial saving of 14 MJ/kg to yield a GER of algal product at 18 MJ/kg, providing a system net energy gain of approximately 5 MJ/kg.

It is claimed that 60 per cent of the algal energy content is recovered as methane [60], so that the mean algal volatile matter energy content of around 23.2 MJ/kg will give rise to 13.9 MJ of recoverable gas. 80 kg of dry weight algae, including ash, are sufficient to produce 1 GJ of gaseous output. At a GER for algae of 18 MJ/kg, the algal GER necessary to yield 1 GJ of 'biogas' energy works out at 1.44 GJ, clearly a net energy loss. The energy inputs required for digester stirring and possibly heating in the conversion process have yet to be included and so, for the present at least, the algae produced in these oxidation pond systems would be far better utilized as a protein provider rather than as a methane source.

The utilization of manure, particularly that from farm animals, as a methanogenic substrate is described in Section 4.2, with some reference to its diverse circumstances of operation made in Section 3.2. Prevailing climatic conditions greatly affect any energy analysis of the process, together with its level of intensity – a familiar recurring theme.

At the low intensity end of the scale a feasibility study has been conducted on a simple Gobar-type anaerobic digestion plant at a small village in Gujarat, India [136]. The digester has no motor-driven parts, and the

Table 5.5 Inputs for practical production of methane from livestock waste via anaerobic digestion in a temperate climate [93].

Major physical inputs	*kg/GJ 'biogas'*	*GER: MJ/GJ 'biogas'*
Cow manure (dry weight)	380.0	0
Water (part of slurry)	4000.0	0
Fuel oil (for digester heating)	18.5	856.6
Electricity (total process)	104.1 kWh(e)/GJ 'biogas'	1557.4
Mild steel (amortized 20 yrs)	3 kg/GJ 'biogas'	150.0
Cement (amortized 20 yrs)	1 kg/GJ 'biogas'	7.8
	Total GER 'biogas' =	2.572 GJ/GJ

amortized capital equipment constituents of mild steel, asbestos-cement, and bricks over 20 years results in a GER of 77 GJ/year. The annual energy content of the 'biogas' output is 156 GJ, giving a net energy gain of 79 GJ/year, and as this particular village site occupies 11.2 ha the energy yield is just 7 GJ/ha yr. This is the equivalent of something just over one barrel of oil per ha yr. This may seem a modest figure, but it does represent a genuinely positive energy return, which is not always the case in more intensive systems where fossil fuel inputs substitute for manual labour and where the digester has to be heated (normally by a proportion of the produced 'biogas') in regions of insufficiently high ambient temperatures.

By contrast, in a more intensive system, where electricity is required as a major input for the agitation of digester contents and pumping etc., the overall energy balance becomes less favourable. A net energy deficit may actually result in such circumstances, but within an integrated system of waste treatment, fertilizer upgrading, and generally sagacious agronomics,the methane gas output may be regarded as something of a bonus and the system operation still worth-while. Gas surplus to the requirements of digester heating is invariably obtained, which can then be used for space heating or even cooking. Table 5.5 shows data relevant to the energy analysis of a UK operation whereby manure from 150 cattle is digested during the six warmest months of the year to give rise to 32 GJ of methane [93], with a total energy input requirement (GER) of 82.5 GJ. It should be stressed that modifications leading to a positive net energy situation have since been proposed using conventional stirred vessels operating at 35 °C and with an ambient temperature of 10 °C, whereby, approximately 1.8 units of 'biogas' energy output are obtained from each unit of renewable energy input [137]. Further innovations are envisaged utilizing a specialized tubular reactor and realizing gas yields of up to 12 fermenter volumes per day, reflecting significant increases in the rates of reaction attained. An independent US study has calculated a more conservative return of 1.32 units of energy from each unit of energy invested [138], but takes into account capital and transportation energy requirements, which were not included in the previous analysis.

However, the analysis of an actually operating mid-1970's system in temperate, energy-intensive Northern Europe presented in Table 5.5 contrasts remarkably with the Indian scenario mentioned above. The tabulated data shows that 40 per cent of the total direct and indirect energy inputs are recovered as 'biogas' energy content. If sufficient waste heat, for instance emanating from a diesel engine, were available, then 60 per cent plus of the input energy could be recouped. As it stands, the pre-optimized operation, requiring a land area of 60 ha, represents a net energy loss of 0.5 GJ/ha yr.

Of the electricity consumption 81 per cent was for digester agitation, and only 19 per cent for substrate preparation and pumping. In the improved system the energy debit for the total digestion process had fallen to 54 per cent, and the energy 'payback' time had dropped from infinity to $\frac{1}{2}$–3 years according to process conditions. The future of such systems, both from the energy and integration viewpoints, looks particularly bright, and a similarly optimistic outlook can be envisaged for fuel gas production from municipal and manufacturing solid organic wastes, which forms the next topic of investigation. Clearly though both technology and management play key roles.

The US Dynatech R/D Company's bioconversion of solid waste to methane was described in Section 4.2. Table 5.6 presents an energy analysis of their computer model-developed process based on the minimization of the unit cost of gas produced as the optimization criterion. The main inputs are electricity for pumping and stirring, and steam for heating. Steam may be produced by utilizing any convenient fossil fuel. Unfortunately in the case analysed, the methane gas formed by the process was not used. At 85 per cent thermal energy recovery and 30 per cent overall conversion of heat to mechanical energy, the ratio of energy required to drive the process (GER) to the energy (gross) generated is a very favourable 0.335. The contribution of chemical inputs such as glycol and monoethanolamine (which are in any event regenerated and recirculated), lime, ferrous salts, and occasionally fresh water is considered insignificant in energy terms. Amortized capital equipment and constructional materials are estimated to have a GER/unit gas output proportional to the operational energy input GER/unit gas output of approximately 1 : 20, and thus be on a par with

Table 5.6 Inputs per 4 TJ methane/24 hours conceptual digestion process from municipal solid waste (after [99]).

Major physical inputs	*kg/GJ methane*	*GER: MJ/GJ methane*
Steam		
(for digester heating)	13.5	45.8
(gas scrubbing)	30.7	104.4
Electricity	13.27 kWh(e)/GJ methane	
(for shredding)	3.76	52.6
(magnetic separation)	0.07	1.0
(screening)	0.04	0.5
(air classification)	1.22	17.1
(2nd shredding)	2.05	28.7
(mixing)	0.86	12.0
(digester agitation)	2.36	35.1
(dewatering)	2.09	29.3
(pumping)	0.11	1.6
(gas scrubbing)	0.71	10.0
Amortized capital energy inputs		16.8
	Total GER =	352.9 MJ/GJ methane

The inclusion of the estimated GER of the capital equipment lowers the ratio of energy out to energy in slightly, but at around 3 : 1 this remains very healthy, and it is to be hoped that such optimistic figures can be realized in practice and not just within the computer's capability.

somewhat more conventional waste treatment operations. The total cost-connected power is 4274 hp, but the time-averaged power consumption is 2916 hp due to only partial use of some items of machinery [99].

5.5 Ethanol via yeast fermentation

Industrial ethanol production from a variety of carbohydrate raw materials has been described in Section 4.3. In essence, the energy gains or losses are dependent upon the extent of pre-treatment required to render the initial substance fermentable by a suitable strain of *Saccharomyces* yeast. The energy intensity of the pre-treatment increases with molecular complexity through sugar cane, cassava starch, the cellulosic materials such as straw, newsprint and wood, the last of which also contains a high proportion of bound lignin. The final GER of the ethanol also depends on whether combustible materials which are themselves biomass constituents, like bagasse or timber, are used to substitute partially for external fossil fuel energy inputs in the running of the process.

Table 5.7 Inputs per kg ethanol derived from sugar cane

Major physical inputs	*kg/kg ethanol*	*GER: MJ/kg ethanol*
Fermentable sugar	2.2	4.89
Bagasse	1.9	2.38
Urea	0.006	0.22
O-phosphoric acid	0.005	0.03
$MgSO_4$	0.001	0.01
H_2SO_4	0.01	0
Antifoam	0.01	0.34
Water	125	0.3
Stainless steel	0.003	0.2
Structural steel	0.004	0.2
Cement	0.008	0.06
Fuel oil	0.18	0.8
Electricity	0.5 kWh(e)/ ethanol	7.0

Total GER ethanol = 24 MJ/kg

Calorific value (gross) of ethanol = 29.7 MJ/kg

Net energy = 97 GJ/ha yr

Table 5.7 shows the input data for one, among several possible, schemes for the ethanolic fermentation of sugar cane followed by distillation to provide the required product at a yield of 17 t/ha yr.

The bagasse forms the combustible, fibrous portion of the sugar cane plant and can substitute partially for the fuel oil requirement, which would otherwise be 0.48 kg/kg ethanol formed, raising the overall GER to 36 MJ/kg. Data presented in Table 5.7 is that of a feasible continuous alcoholic fermentation, followed by yeast cell separation, and distillation of the spent medium. Continuous culture is more rapid and gives slightly higher yields than does batch, but necessitates thorough sterilization procedures since the problem of contamination is magnified. This is reflected in the higher proportion of sterilization steam over distillation steam (23 per cent : 77 per cent) [11] than would be required in a batch operation (9 per cent : 91 per cent). A more conventional non-continuous batch fermentation involving molasses as substrate with ammonium sulphate as additional nitrogen source would have a GER for ethanol of around 28.30 MJ/kg, which could be lowered to less than 25 MJ/kg were bagasse available as an energy input. It would also be possible to utilize both the saccharide and fibrous portions of the sugar cane plant as ethanolic substrates, but with an increase in GER of the ethanol to approximately 97 MJ/kg. Partially offsetting this increase in energy intensity is a decrease in land area requirement of around 30 per cent, but this would hardly be sufficient compensation for turning the process into a clear net energy sink. Finally, referring back to Table 5.7, it may be observed that the sulphuric acid input has been allocated a zero energy cost. This arises in view of the wide variety of methods adopted in its production, and the fact that only an integrated plant can utilize the energy surplus within its production process.

Dealing with cassava growth, the preparation of amylase enzymes, the enzymatic hydrolysis of the starchy cassava tubers to sugars, and the final microbial fermentation and distillation are as described in Section 4.3. Table 5.8 presents the overall operational inputs. Taking into account the land area requirement for substrate growth, the ethanol yield is approximately 2 t/ha yr.

When energy analyses of the individual operations within the total process were carried out and summed then the GER for ethanol was 59 MJ/kg. A value of 60 MG/kg is therefore taken as being reasonable, giving a net energy of: 30.3 MJ/kg.

For every kg of cassava tubers produced there is also 0.7 kg of tops which could be combusted as a fuel, though this fibre source is considerably less in quantity compared to sugar cane bagasse, and also contains a higher proportion of water. As stated in Section 3.1, concerning the Brazilian national alcohol programme, it is debatable whether the cassava-to-ethanol route is

Table 5.8 Inputs per kg ethanol derived from cassava starch

Major physical inputs	*kg/kg ethanol*	*GER: MJ/kg ethanol*
Cassava starch (solubilized)	2.45	19.19
Cornsteep liquor	0.01	0.06
Urea	0.006	0.22
O-phosphoric acid	0.005	0.03
$MgSO_4$	0.001	0.01
H_2SO_4	0.12	0
NaH_2PO_4	0.012	0.171
$CaCl_2 2H_2O$	0.0003	0.001
$MgCl_2 6H_2O$	0.0005	0.003
KCl	0.0001	0.001
Antifoam	0.0115	0.39
Water	159	0.38
Stainless steel	0.0055	0.374
Structural steel	0.0117	0.585
Cement	0.032	0.253
Fuel oil	0.627	29.03
Electricity	0.784 kWh(e)/kg ethanol	10.47
	Total GER ethanol =	61 MJ/kg
	Net energy =	– 31.3 MJ/kg

indeed a net energy producer, although the above analysis would indicate that it is not, since the calorific value of pure ethanol is 29.7 MJ/kg. A further appraisal [139] also suggests that the overall conversion entails a net energy deficit, while a third [128] claims the reverse, but with the proviso that the cassava cellulose tops can be burnt to provide most of the process steam, a premise about which there is much doubt [18, 140]. Nevertheless, it might just be possible to cut the fuel oil requirement in the region of 50 per cent to make the process feasible because of the thermodynamically high availability of the product, and the Brazilian assurances mentioned earlier [20, 21, 132] do indicate that positive net energy returns are already occurring.

Continuing finally to the cellulosic and ligno-cellulosic raw materials such as newsprint, straw and timber, Tables 5.9 and 5.10 respectively contain the required inputs for two wood-ethanol schemes, the former involving acid hydrolysis to sugars and subsequent alcoholic fermentation and distillation, and the latter substituting enzymatic hydrolysis, both as described in Section 4.3. On the basis of ethanol yields from land which is annually renewed with the quantity of timber consumed in the process, these yields become around 0.2 t/ha yr and 0.35 t/ha yr respectively. The acid hydrolytic procedure was based on ideal operation at the German

Table 5.9 Inputs per kg ethanol derived from timber (using acid hydrolysis)

Major physical inputs	*kg/kg ethanol*	*GER: MJ/kg ethanol*
Wood	6.25	20.0
Superphosphate	0.005	0.07
Na_3PO_4	0.005	0.07
$(NH_4)_2SO_4$	0.008	0.12
CaO	0.28	2.52
$CaCO_3$	0.38	3.42
H_2SO_4	0.42	0
Antifoam	0.005	0.17
Hard coal	2.08	62.4
Stainless steel	0.004	0.26
Structural steel	0.006	0.3
Cement	0.01	0.08
Water	125	0.3
Electricity	0.56 kWh(e)/kg ethanol	7.84

Total GER ethanol = 98 MJ/kg

Calorific value (gross) of ethanol = 29.7 MJ/kg

∴ Net energy = –68.3 MJ/kg

Table 5.10 Inputs per kg ethanol derived from timber (using enzymatic hydrolysis)

Major physical inputs	*kg/kg ethanol*	*GER: MJ/kg ethanol*
Wood	3.96	12.67
Cellulose	0.24	0.15
Urea	0.021	0.76
Peptone	0.038	0.76
O-phosphoric acid	0.005	0.03
$MgSO_4$	0.009	0.08
$CaCl_2$	0.015	0.02
KH_2PO_4	0.1	1.4
$(NH_4)_2SO_4$	0.07	1.02
H_2SO_4	0.01	0
Antifoam	0.014	0.48
Fuel oil	0.91	42.13
Stainless steel	0.017	1.16
Structural steel	0.028	1.4
Cement	0.1	0.78
Water	320	0.8
Electricity	12.55 kWh(e)/kg ethanol	175.7
	Total GER ethanol =	239 MJ/kg
	Calorific value (gross) of ethanol =	29.7 MJ/kg
	∴ Net energy =	−209.3 MJ/kg

Holzminden plant where 3500 t of ethanol could be produced annually during and after the Second World War [14].

This process is also a net energy loss. If wood were substituted for coal then the overall GER for ethanol could be reduced significantly, possibly to 50 MJ/kg, still a net energy loss, and at the same time increasing the land use requirement by 100 per cent or more.

The wide discrepancies in the energy inputs to the two wood-to-ethanol processes are brought about firstly by the large electricity requirement for the preliminary ball-milling treatment of the raw material in the enzymatic conversion, and secondly by the greater inputs necessary for the two stages of enzyme induction and hydrolysis over the single acid hydrolysis stage. Conversely, the acid route is more labour-intensive. In both cases the GER for wood includes energy expended in chopping and initial size reduction.

It is just conceivable that wood could be used as the sole primary energy source and thereby reduce the GER for ethanol to about 70 MJ/kg via the enzymatic route. This would still represent an overall net energy loss, and at the same time raise the land area requirement by a factor of five. In many parts of the world the problems associated with deforestation are so acute that such an increase could hardly be permissible.

The GERs for straw- and newsprint-derived ethanol via the enzymatic route are slightly less than that from wood, having been calculated at 222 MJ/kg and 225 MJ/kg respectively, due principally to the lower GERs of the raw materials themselves [131]. Once again, if fossil fuel inputs were wholly or partially replaced by combustible biomass, then the GERs would fall at the expense of greatly increasing the land area requirement, probably to an unacceptable degree in many countries [141].

In the case of these cellulosic and ligno-cellulosic raw materials, ball-milling, an energy-intensive procedure, appears to be by far the most effective initial treatment for releasing the cellulose molecules from the binding ligno-cellulosic complex, and for decreasing the particle size and reducing the crystallinity of the cellulose substrate to a form amenable to enzymatic attack. It has been claimed (see Section 4.3.5) that hammer-milling, at only 3.6 per cent the electricity requirement for ball-milling, is adequate for a satisfactory conversion [114], but experience at the US Army Natick Development Center, Mass., has demonstrated that quite extensive ball-milling is necessary, to the extent of making it the single greatest cost factor in the economic evaluation of the enzymatic processing of cellulosic raw materials [113, 117]. It is to be hoped that ball-milling may be dispensed with, nonetheless, in the not-too-distant future, owing to the development of the ligno-cellulose degrading enzyme complex, produced by the white rot fungus, *Sporotrichum pulverulentum* in Sweden (see Section 7.6). Without the prospect of such advances to reduce energy consumption, then the direct combustion of substances like wood would appear to be energetically more productive than their conversion to alcohol.

The estimated values of the GER/kg ethanol produced from the above biomass materials using intensive Western technology are shown in Table 5.11, with a further breakdown into requirement categories.

In an integrated operation the majority of the yeast by-product formed in each of the final fermentation stages, about the equivalent of 0.07 kg yeast protein/kg ethanol, could be sold as an animal feedstuff. Since yeast protein grown on a molasses substrate has a GER of 75 MJ/kg [142], an energy credit of 5.3 MJ/kg could accrue. Additionally, alcohol-tolerant yeast strains are being bred to withstand high sugar concentrations and which ferment to around 12 per cent v/v ethanol, so that the increased ethanol concentration passing to the still allows for a 40 per cent reduction in distillation steam [11]. This would mean a decrease of 0.15 kg fuel oil/kg ethanol, and an energy saving of 6.9 MJ/kg ethanol. Combining the two, there is the immediate possibility of lowering the GERs of Table 5.10 by 12 MJ/kg making the sugar cane-ethanol route a particularly attractive net energy winner.

Table 5.11 Gross energy requirement (MJ/kg) of ethanol production from biomass substrates [131].

Physical inputs	*Substrate sugar cane*	*Substrate cassava*	*Substrate timber*[a]	*Substrate timber*[b]	*Substrate straw*
Substrate	7.27[c]	19.19[d]	12.67	20.0	4.37
Additional chemicals	0.6	0.89	4.74	6.37	4.74
Water	0.3	0.38	0.8	0.3	0.8
Electricity	7.0	10.47	175.7	7.84	166.74
Fuel Oil	8.0	29.03	42.13	62.4[e]	42.13
Capital inputs (e.g. steel and cement)	0.46	1.21	3.34	0.64	3.34
Total	24	61	239	98	222
System net energy/ha yr	97GJ	negative	negative	negative	negative

[a]Fermentable sugars formed via enzymatic hydrolysis

[b]Fermentable sugars formed via acid hydrolysis

[c]Value includes that for sugar cane bagasse as well as for crude sugar

[d]Includes energy for solubilization

[e]Hard coal was used in place of fuel oil

Calculations are derived from the indicated references, in part, and from the normal requirements of ethanolic fermentations.

As for the wood to alcohol route, substantial technological development is called for before it can be considered a net energy producer. There is, however, another argument that can be applied, namely the one used for the production of food. We do not only consider food in terms of its calorific value, but in terms of its metabolizable calorific value and nutritional value. In a similar way biomass is simply a source of heat, whereas when converted to ethanol, it can be used in a sophisticated device such as an internal combustion engine. Even so, the net energy penalty on existing wood or straw to ethanol processes is so great, that it is, for the moment, more attractive to return to the age-old process of wood distillation.

5.6 A miscellany of fuels from biomass systems

It has been necessary to concentrate upon energy analyses of anaerobic digestion to methane and microbial fermentation to ethanol simply because there is a distinct lack of adequate data for other bioconversion routes. Even where independent attempts have been made on essentially identical processes different conclusions may be arrived at owing to different assumptions made or differing interpretations drawn. Thus, the debate touched on in Section 5.5 as to whether or not ethanol production from cassava is or is not a net energy provider, an issue which could be crucial in the context of the Brazilians' desire to utilize the plant as an energy source on a quite massive scale.

A number of interesting studies have been carried out in Australia, a country with much potential for biomass energy, which draw comparisions between various fuel production scenarios on the basis of both energy and financial budgets. The economic aspects will be deferred until Chapter 6, with the energetics of seven conceivable processes presented in Table 5.12 below.

It is immediately apparent from Table 5.12 that the considerably higher energy inputs into a cellulose-based alcohol production process over a starch-based operation render the former a definite net energy loser. The energy balances in all cases might well be improved by the development of a continuous hydrolysis/fermentation process. The most favourable net energy occurs using the pyrolysis technique on *Eucalyptus*, but ethanol has the advantage over pyrolytic oil in that it can be blended with petrol in conventional internal combustion engines. Comparing pyrolysis and

Table 5.12 Energy costs and efficiencies of photobiological fuels (adapted from [128], [130])

Fuel	*Raw material*	*Process*	*GER (MJ/kg product)*	*Net energy (GJ/ha yr)*
Alcohol	Cassava tops* and tubers	Enzyme hydrolysis/ Batch fermentation	17.3	+ 80
Alcohol	*Eucalyptus*	Acid hydrolysis/ Batch fermentation	105.0	−452
Alcohol	*Eucalyptus*	Enzyme hydrolysis/ Batch fermentation	>105.0	<−452
Methane	Cereal straw	Anaerobic digestion	20.0	+ 7
Methane	*Eucalyptus*	Anaerobic digestion	20.0	+ 84
Pyrolytic oil/Char	Cereal straw	Flash pyrolysis	4.8	+ 11
Pyrolytic oil/Char	*Eucalyptus*	Flash pyrolysis	4.8	+131

*It is assumed that the cassava cellulose tops can be burnt to provide most of the process steam.

anaerobic digestion to methane from cereal straw, it is concluded that pyrolysis is perhaps the more attractive and may be feasible in the short term provided a secure, high value market could be discovered for the char in addition to that for the produced pyrolytic oil, which may be utilized as a No.6 fuel oil extender. It has a lower sulphur content than standard No.6 fuel oil, but a calorific value of only 23 MJ/kg as against 42.25 MJ/kg [129].

A particularly optimistic assessment has been made regarding the growth of *Eucalyptus* forest plantations within Australia and utilizing the wood yield to produce ethanol via acid hydrolysis and yeast fermentation. It is estimated that 13 million hectares, 50 per cent of the arable land not already cultivated, could meet half the nation's requirement for transportation fuel by the year 2000 (approximately 1 EJ/year). In the process envisaged 20 per cent of the wood energy content is transformed into liquid fuel energy, with most of the remaining 80 per cent utilized to provide process steam [143]. It is claimed that three times more fuel would be obtained than that consumed in the overall operation, but it is unclear whether indirect material energy inputs have been included. Even if they have not there would still be a remarkably high net energy yield, although another study maintains a significant increase in land area requirement owing partly to *Eucalyptus* productivity being only around 50 per cent of that claimed in the first investigation, and to the recovery of only 13 per cent wood energy content in the ethanol product, rather than the 20 per cent maintained above [139].

Finally, a similarly high net energy yield is claimed for the Occidental Research Corporation's flash pyrolysis process described in Section 4.6 [122]. Each tonne of municipal refuse requires the following quantitites of electricity for the various processing stages:

Process stage	*Electricity consumption (kWh)*
Preparation for Pyrolysis Reactor	60.0
Aluminium Separation	7.2
Glass Separation	2.6
Pyrolysis Reaction	63.6
Afterburner and Utilities	6.0
TOTAL	139.4

No supplemental fuel input is required, and the amortized capital energy inputs may be assigned a GER/unit pyrolytic oil energy output proportional to the operational energy inputs of around 1 : 20. This was the convention given in the Dynatech R/D Company's methane production from solid waste process [99]. The electrical consumption totals 139.4 kWh(e) giving a GER of 1952 MJ. Adding the GER of the capital inputs a figure of 2049 MJ/t refuse processed is obtained. The energy content of the pyrolytic oil obtained is 5430 MJ (in 197 litres, of which 14 per cent is water), to give a net energy credit of 3381 MJ/t refuse. Additionally, if 20 per cent of the pyrolytic char were also available as a product the net energy gain would be near 3516 MJ/t refuse, and this does not include energy credits for the recovery of iron, aluminium and glass in the pre-pyrolysis process. The process is thus entirely energy self-sufficient, with an 'energy out : fossil fuel energy in' ratio of approximately 2.7 : 1, assuming a GER of municipal refuse of 0, though a negative GER could rationally be applied here, if known.

Patently, further energy analyses are required on systems such as these, but with clearly defined premises and conditions. There is also a great need for much more data collection from actual operational processes rather than paper study interpretations alone, but equally it would amount to crass folly to invest huge quantities of labour, materials, and money into a proposed scheme without firstly attempting some kind of energy analysis. At least then it might be justifiable to close the stable door after the horse has bolted, as long as it is retrieved before jumping the farm gate. In other words a preliminary energy analysis may not provide a conclusive answer either way, but even if it subsequently transpires that a particular scheme will not be a winner, sufficient knowledge will have been gained to enable effective transference of the available resources without all being lost. It must be remembered that energy analysis is a very useful tool used properly, but does not pretend to be the 'be-all' and 'end-all' of any decision-making process. It has its limitations, just as purely economic approaches to resource questions have, and it is hoped that the cost-benefit analyses and energy analyses introduced together in the next chapter will be seen rather to complement each other than detract from each other.

Chapter 6

The economics of biomass systems

6.1 General considerations

It is frequently argued in certain quarters that, because solar energy is a free good, the economics of solar energy devices are bound to be favourable. This would certainly be true, if land had no value, and capital had no cost. Land, however, forms a basic part of bioenergy systems, because the capture of solar energy is a function of the surface area provided. The harvesting of bioenergy from natural ecosystems can be done without capital, in the sense of capital for technology, but the intensity of production is so low that such systems can only satisfy their immediately adjacent inhabitants, and are therefore restricted to highly dispersed systems or to comparatively primitive ways of life.

On the face of it the economic evaluation of a bioenergy system should be comparatively simple. One is, after all, comparing different routes to the same product, namely a Joule of heat or a Joule of work, each of which has the same function whatever its source. It is a commonly held view that the only thing needed to make solar energy systems economic is a rise in the price of its competitors, that is to say, of usable energy derived from fossil or fissile sources. This can only be true for unintensified natural bioenergy systems where the land cost is negligible; and where may one find such situations in the world today? Perhaps in the inner jungles of the Amazon. Yet the indigenous inhabitants of those lands lack the technology with which to exploit their virtually free fuel supply, while the incoming exploiter sees a market for the bioenergy product as a material of construction rather than as a fuel, and he sells it in those terms upon the world market. The proposition is fundamentally untrue, as soon as one starts to deal with intensified bioenergy systems for the following reason. The cost of the energy coming from an intensive bioenergy system has to be costed, not against the cost of fuel from fissile or fossil sources but through the cost of energy from those sources, since energy is required to inaugurate the bioenergy systems themselves. For a detailed argument around this point the reader may care to consult Slesser, *Energy Analysis, Uses and Limits* (RM-78-46, International Institute for Applied Systems Analysis, Austria, 1978). What it boils down to is this: When a change in the price of energy delivered to our economy occurs through market forces, that price increase percolates throughout the entire system, raising prices of manufacturing processes, raw materials, the price demanded by labour, the cost of running homes, transport and so on. The intensification of a bioenergy process requires investment in capital structures and the acquisition of many inputs. A rise in the price of energy simply raises all these inputs at a rate which in turn is determined as much as anything by their energy intensity. Thus, after a round of energy price increases from fissile or fossil sources, we find that the cost of running the bioenergy systems has also risen, and that there is little or no gain. There is, however, one time-window during which one can benefit from a price increase, and that is immediately following or immediately anticipating a general fuel price increase, for at this point the capital costs are still at one level, and the bioenergy system may be developed using those costs, whereas by the time the system is operational, the energy cost of the competitor, the fossil or fissile energy, will have risen. For example, those people who in October 1973 immediately invested in double glazing of their houses were able to do so at a cost which today seems comparatively trivial and which brings considerable financial benefits in relation to the capital expended. Now that energy prices have to some extent stabilized again, the economics of double glazing are nothing like as favourable. What follows from this reasoning is that bioenergy systems have to be intrinsically 'economic', if they are to succeed. Here, however, we face exactly the same problem in assessing their economic viability as we would with any new technological process, using fissile or fossil fuel, namely the high degree of uncertainty surrounding future fuel prices. Table 6.1 lists some of the price changes that occurred between December 1970 and March 1974 in the UK. The changes shown, though large, are as nothing compared to the subsequent changes between March 1974 and December 1976.

Economics is by definition the science of scarcity and therefore is con-

Table 6.1 Fuel prices to domestic consumers (UK)

		December 1970 [*144*]		*March 1974* [*144*]		*December 1976* [*145*]	
Coke	p/kg	2.03	(100)	2.68	(132)		
Anthracite	p/kg	2.09	(100)	2.47	(118)	4.64	(222)
House coal	p/kg	1.45	(100)	1.87	(129)	3.05	(210)
Fuel oil	p/l	2.04	(100)	4.71	(231)	7.68	(376)
Paraffin oil	p/l	2.95	(100)	6.16	(209)	9.57	(324)
Gas	p/kWh	0.37	(100)	0.41	(111)	0.57	(154)
Electricity: on peak	p/kWh	0.78	(100)	0.95	(122)	2.36	(303)
Electricity: off peak	p/kWh	0.35	(100)	0.43	(123)	1.09	(311)
RETAIL PRICE INDEX		100		134		219	

(x) = indices

cerned with making decisions about the best use of resources. Within conventional economic theory an investor in the market place will examine the opportunities available to him, and choose that one which will give him the most satisfactory return on his investment. Hence a proposed bioenergy system would be examined in that context. The amount of knowledge one must have about the future to make such a decision is greater than the information available. Therefore an element of uncertainty, that is to say, of gamble, enters into the computation. One of the elements of this gamble is to ascertain or guess the appropriate discount rate. Now the discount rate is a number representing an anticipated average return on investment in the future.

A commonly adopted rate is 10 per cent, often used in national accounting as a rate for testing the 'viability' of government-supported activities. Of course, all entrepreneurs hope for a rate of return in excess of this. Some are justified in their hopes and some are very disappointed. The economic approach to viability is to argue that a pound in your pocket today is worth more than a pound in your pocket in one year's time, and that, if the discount rate be taken as 10 per cent, then in order to be as well off next year as now, if you are not going to receive a pound today, you must be promised one pound + 10 per cent = 1.1 pounds one year hence, £1.33 four years hence and so on. Moreover, £1.33 of profit four years hence is equivalent to one pound now. Hence, if an investment of say £10 000 is required to set in train a bioenergy system, the future profits of that bioenergy system must be discounted according to how many years ahead they occur. Under the 10 per cent discount rule, if the accumulated profits discounted back to the day of investment do not exceed the investment, then the investment is judged to be economically unsound. The weakness of this approach may be revealed in the fact that though for all normal activities supported by Government money in the UK, the Treasury demands that a 10 per cent discount rate be applied, it was forced to abandon this value when it came to forestry plantations. Under the 10 per cent rule it was simply not economically viable to grow forests, yet it was plainly obvious the trees had to be grown and, moreover, were worth growing.

In fact, a test discount rate of 10 per cent is no more than what it says, that is to say, a test. It has been used as a yardstick for deciding whether this or that investment has merit or not, and removes from the person making the calculation the need to make a decision as to whether he should go ahead or not. In fact, a more realistic measure is to establish the internal rate of return of the investment. Both these items will be explained in greater detail in the succeeding section. For the moment, let us consider the effect of the rate of change of energy prices on the economic viability, using conventional discounting techniques. Let us take a situation where an investment of £1 million yields an initial saving of £100 000. Below we show the number of years of operation needed to realize a net economic return on this investment for rates of return of 10 per cent and 2 per cent.

Rate of rise in energy price/yr (%)	0	3	6	10	15
Years to achieve viability at 10% IRR	never	18	13	9½	8¼
Years to achieve viability at 2% IRR	25	9½	8½	7¼	6¼

At the heart of the problem is the complete uncertainty about future energy prices, and the lack of formulation in the economic literature on the impact of energy prices on all other prices. This uncertainty has encouraged the development of another way of looking at the viability of energy systems, whether they be bioenergy systems or not. This is the concept of energy payback time. To give some examples: at a time when photovoltaic solar cells made from cut silicon were costing $20 000/kW peak power output, one could show by energy analysis that over the lifetime of a cell its cumulative energy output would be substantially less than the energy required for its manufacture. This was a case where the point of futility had been passed, but was acceptable because of the unique use to which solar cells were put to at that time, namely into space capsules. Today, when photovoltaic cells are approaching $3000 a peak kW, the energy payback time is of the order of 1–2 years. That is to say, that within one or two years of commencing operation, the total energy output of the system will have been as great as the energy required to build the system. Thereafter energy flows free of charge. Such calculations are more difficult to make for bioenergy systems because of the diverse nature of these systems and the diverse locations where they may be found, varying from intensive sewage treatment plants in industrialized countries to simple 'biogas' generation in Third World countries. The relationship between energy payback time and economic viability is considered in Section 6.3.

6.2 Net present value (NPV)

The future savings of a particular process, such as a bioenergy system, may be discounted back to the day of the investment. This, in number, is called the present value (PV). The difference between this number and the capital investment is therefore the net present value (NPV).

The formula by which one can establish the present value (PV) of a future activity is:

$$PV = \frac{S(1 + r)^t - 1}{r(1 + r)^t}$$

Where S = a uniform, annual sum, such as an anticipated saving
t = number of years
r = selected discount (interest) rate

The means of computing this number will be found in Appendix 3.

Let us now examine the application of this formula in the context of a solar device, namely a 4 m square solar energy collector which is expected to provide 4.8 GJ of energy each year for water heating. A capital cost installed of £350 is estimated. Water can also be heated by electricity from the public supply, which may be obtained at 2.7 pence per kWh electric, and we will ask the question whether it is cost-effective to purchase the solar hot water collector. To answer this question we need to make a number of assumptions: (1) that the price of electricity will remain unchanged through time; (2) that the temperature of the water in the solar water heater is high enough to meet the demand put upon it; (3) that we know the life of the device, say 15 years; (4) that we know the likely rate of return on alternative possibilities for investment; this is described as the test discount rate and is frequently taken as 10 per cent (or 0.1).

The saving is the annual cost of electricity no longer needing to be purchased – that is, the heat equivalent of 4.8 GJ (or 1333 kWh) which, at 2.7 p/kWh costs £36.00. This is the sum saved.

$$S = £36$$

Hence: $$PV = \frac{36\left(\frac{1}{1+0.1}\right)\left\{1-\left(\frac{1}{1+0.1}\right)^{15}\right\}}{1-\left(\frac{1}{1+0.1}\right)}$$

$$= £273.08.$$

But the actual initial expenditure is £350.

We see that the present value of the solar collector is £273.08, whereas the capital investment was £350. Hence the net present value is −£76.92. It would therefore be argued that the solar energy system was not 'cost-effective'. Before exploring further the implications of this statement, we can also carry out another calculation, in which we establish what is the discount rate which will render the device cost-effective, that is to say, which will produce a net present value exactly equal to the original investment. In economic language this is referred to as the internal rate of return (IRR); this works out at approximately 6 per cent.

This number is obtained by setting the *PV* to the initial capital outlay, and then finding the value of r which satisfies this condition. Taking the above example:

$$PV = £350$$

$$\text{Annual savings} = £36$$

$$\therefore \text{ the relation } \quad \frac{(1+r)^t - 1}{r(1+r)^t} = \frac{350}{36} = 9.72$$

The equation is now solved for r, most readily done using the information given in the Appendix 3 tables, where the internal rate of return for a 15 year lifespan is 6 per cent.

Now, before we utilize this formula to either accept or reject the construction of a solar energy device, we have to look very closely at the four assumptions. Of these, two are indeterminant; the life of the collector and the future price of the alternative energy for heating the water. We may be able to judge the first by experience, which may tell us, for example, that collectors usually last about 15 years. So that is acceptable as an assumption. One should note that optimists 'pushing' a certain technology, can try to improve the NPV by assuming a longer lifespan, and frequently do. But we are in a much more difficult situation with respect to the future price of alternative energy. In the formulae one sees that the annual saving is noted as £36 per annum, which is obtained by computing the cost of the alternative fuel – in this example, electricity. Suppose, however, that the cost of electricity were to rise 5 per cent a year then clearly the net present value at the same test discount rate would be substantially higher. Thus this calculation of net present value and cost-effectiveness, while intellectually very satisfying is of little value in practical terms unless one can predict future prices. The NPV formula can be changed to allow for this in the form of a general inflation factor. Let us express this as i, an annual (fixed) percentage. Then:

$$PV = \frac{S\left(\frac{1+i}{1+r}\right)\left\{1-\left(\frac{1+i}{1+r}\right)^{t}\right\}}{1-\left(\frac{1+i}{1+r}\right)}$$

Values for $i = 0$, 3 per cent, and 10 per cent are given in Appendix 3.

Let us suppose we have advice which suggests a 5 per cent annual rise in the price of electricity. Then $i = 0.05$, and we have $PV = £380$ and $NPV = £380 - £350 = £30$. That is to say the device *is* cost-effective. It is clear that this sort of calculation is heavily predicated on three unknowns:

1. The prevailing interest rate – effectively the discount rate, r.
2. The rate of rise of energy prices.
3. The expected life of the system.

In these circumstances it is not difficult to manipulate the figures to give the desired result. The method must be used with considerable caution.

One way around this problem is to look for the link between net energy/hectare-year and cost. Consider a steady-state biomass system having a total land requirement of h hectares, whose rent is £r/year. The system is furnished with inputs from the economic system, costing £p/year and has a capital charge of £c/year. It yields a steady-state output of g GJ of energy/year. However the operation consumes an internal recycle of energy, E, so that the net energy, $y = g - E$ GJ/year.

The output cost, G = (rent + capital charges + inputs)/y

$$= (r.h + c + p)/y/\text{year} \qquad (1)$$

The net energy per unit area, N, is:

$$\frac{\text{output energy} - \text{input energy}}{\text{area}}$$

$$N = \underline{[g - E - \{c \cdot I_c + pI_p\}]/h} \qquad \text{GJ/ha/year}$$

$$= \frac{y}{h} - \frac{cI_c}{h} - \frac{pI_p}{h} \qquad (2)$$

Where I_c is the average energy intensity of the capital (GJ/£) and I_p that of inputs.
Rearranging Equations (1) and (2) in terms of h/y, we have

$$\frac{G}{r} - \frac{c}{yr} - \frac{p}{yr} = \frac{1}{N}\left\{1 - \frac{cI_c}{y} + \frac{pI_p}{y}\right\}$$

whence,

$$G = \frac{r}{N}\left\{1 - \frac{1}{y}(cI_c + pI_p)\right\} + \frac{1}{y}(c + p)$$

That is to say:

$$G = \frac{a}{N} + b \qquad (3)$$

where,

$$b = \frac{1}{y}(c + p)$$

$$a = r\left\{1 - \frac{1}{y}(cI_c + pI_p)\right\} = r(1 - bI)$$

It is seen that cost, G, falls as N increases, to a limiting cost $(c + p)/y$. For a natural ecosystem, hand-harvested, $(c + p)/y$ = zero.

It is instructive to consider some plausible values for a and b in Equation (3), taken from a biomass plantation study. To simplify we take $I_c = I_p = 0.20$ GJ/£, $N = 30$ GJ/ha year, $c + p =$ £2.4×10^5, $y = 64$ GJ/ha and $h = 3125$ hectares.

Then

$$b = \frac{2.4 \times 10^5}{3125 \times 64} = \text{£}1.2/\text{GJ}; \quad a = r(1 - 0.2 \times 1.2) = 0.76\,r$$

That is,

$$G = \frac{0.76\,r}{30} + 1.2$$

The roles of land rent and energy drawn from the economy stand out clearly. If there were no energy in the form of goods sucked in from the economy ($= 64 - 30 = 34$ GJ/ha), there would be no costs, $b = 0$ and $a = 3$, so that costs $G = r/64$.

Equation (3) does not define a single line, for N, p and c are all partially dependent and r is quite independent. But one can see the general trend of the relationship r in Table 6.7 where both N and G for several systems are quoted. On current energy costs, a yield of 100 GJ/ha.year appears to be about a break-even cost.

What then can one do to establish the economics of such a system? One way is to establish the internal rate of return, which in the example first quoted will give you 6 per cent, and then simply formulate a mental model about future energy prices. If it is one's hunch that energy prices will rise by more than 2 per cent a year, then lo and behold the collector is cost-effective at an 8 per cent rate. However, if this method has an element of uncertainty about it, then the calculation is largely an academic exercise.

Thus, it is comparatively easy for the advocate of this or that bio-technology to demonstrate that it is 'viable' or 'cost-effective' just as proponents of hard technologies like nuclear energy do. It is less easy to establish the truth of the matter.

6.3 Energy payback time

An alternative approach, which is increasingly being utilized, is to establish the energy payback time. If the rules of energy analysis are strictly followed, this can be an extremely revealing exercise and has demonstrated in its time the non-viability of many a proposed technology. In this approach one does not use money at all; one estimates the fossil or fissile energy required to construct and install a solar device and so determine the number of years of operation before that energy is paid back. Taking the example of our 4 m^2 solar energy collection system, a typical figure for the energy investment for that would be 20 GJ. Hence in 20/4.8 years the 'energy investment' is paid back, that is to say, a payback time of 4.16 years.

Is this good or bad? Before answering, let us examine the logic of the situation. If by an investment of 20 GJ we have a device which will furnish energy free for two centuries at the rate of 4.8 GJ/year, a far-sighted community would certainly regard it as a good investment. In earlier times this attitude was prevalent. Incas built terraces on steep mountain sides, which have yielded a crop of maize over a period of a thousand years. The initial capital input was labour, rather than energy. This initial input was prodigious; the labour involved could have tilled the fields of more accessible land for many a year. Coming to more recent times, water mills were common forms of energy sources a hundred or two years ago. Their construction took both labour (time) and energy. Their output lasted, with maintenance, indefinitely. They were only superseded because intense sources of energy became cheap and technology provided means of bringing energy to bear on exactly the appropriate point of use, obviating the clumsy energy transfer of the old water wheel systems. Today water wheels coupled to modern technology can once again compete.

Thus in evaluating, say, the solar collector, we must set the 20 GJ investment against the return (4.8 GJ) and the life-time. At a 15 year life-time (as in the calculations above), it takes 4.16/15 or 28 per cent of the system's life to pay for itself. How does one then compare this with the hard technologies? It is estimated that a coal-fired power station will last 20 years, and pay back its energy investment in three months. The payback time is 0.25 years and the percentage of the system life is 1.25 per cent. Incomparably better? The problem is we are not comparing like with like. The solar collector is actually producing energy (in the sense of giving an output from a free input that exists whether used or not). The coal-fired power station is not producing energy at all; it is consuming it. To compare like with like, let us compare the current technology of producing electricity from a solar system with that from coal, as in Table 6.2.

Some interesting points emerge from Table 6.2. Firstly, in the true sense a coal-fired power plant never pays back its energy requirement because though it requires only 0.25 years of output for the initial investment, its consumption during operation over its 20 year life adds a further 66 years of output which, by current technology, would be needed to create the input of fossil fuel (this assumes a 33 per cent efficiency of heat to electricity and a GER for coal of 1.1).

Table 6.2 points out the fact that solar energy systems must be looked at in terms of their capacity to create energy, rather than consume, that is to say, their potential to breed.

This concept of energy breeding, called the solar breeder, was first exposed by Slesser and Hounam [146]. They postulated the situation where a solar energy device producing energy uses that energy for three purposes. The first purpose is to supplant it with a replacement device at the end of its life; the second purpose is to contribute to the construction of an additional device; and the third purpose is to sustain the energy system supply of the economy. Clearly, depending on the original investment, the life of the system and the proportion of the output one allocates to developing new systems, the system as a whole can grow at a slower or faster rate. It emerged from this study, that the definition or criterion of success of a solar energy system was the potential doubling time of the system. For example,

Table 6.2 Comparative economic assessments of electricity from renewable and non-renewable sources using current technology, 1977 prices.

	Solar-electric	*Coal-fired power station*
1. Investment per GJ/year output		
money	£800	£25
energy	4 GJ	0.25 GJ
2. System life	25 years	20 years
3. Pay-back time for energy		
capital	4 years	0.25 years
fuel	nil	66 years
4. % of life to pay back		
capital	16%	1.25
capital + fuel	16%	never
5. NPV, assuming 5% fuel inflation, and 10% discount rate	£170	depends on foreseeable profit
6. Cost-effectiveness	−£630/GJ (negative value)	about zero

for a solar breeding system whose original device had a payback time of 10 years, and with a policy of investing 5 per cent of the output in additional devices then this system would have a doubling time of 14.5 years. This is not impressive; nevertheless it is substantially larger than the rate at which energy demand is expanding globally. On the other hand, given substantial improvements in solar technology, so that the payback time fell to two years, and allowing a substantial increase in the fraction allocated to additional devices, one could reach an optimal doubling time of 3.2 years, a quite remarkable rate of enhancement of the solar driven energy supply system, and one which would allow solar driven energy systems to meet global energy demand within a mere 40 years of initiation.

How then can these thoughts be translated into the area of bioenergies? In principle, a bioenergy system is in no sense different from that of a hard solar collecting device like a manufactured solar collector or a photovoltaic device. The points of issue are exactly the same. One must acquire a knowledge of the land area required, the initial energy investment, the time to pay back that initial energy investment, the projected life of the system and last, but by no means least, a clear understanding of how the output energy quality can be transferred into the type of energy required for the construction. Thus, in our view, the energy payback time is not as simple a calculation as was expressed in the preceding pages for there we allowed ourselves to assume that a GJ of output from the solar collector was equivalent to a GJ of input into making it. That ignores the fact that high temperature heat is worth more than low temperature heat. However, that is an energy analysis disciplinary problem and can be handled by evaluating all inputs and outputs in terms of available work rather than in terms of heat. The analysis of the economics and merits of solar energy systems along these lines is very much in its infancy, for there is comparatively little information garnered or published which is at all reliable. Certainly the novelty of this approach will not penetrate the accepted mode of economic thinking for some time, and proponents of bioenergy systems will undoubtedly be obliged to argue the merits of their particular technology on the grounds of conventional economic analysis. A criticism of the energy analysis and payback method is that it puts no value on time, that is to say, there is no discount rate for energy. In this sense the economists argue such an analysis puts the same value on a Joule of energy 20 years hence as today; it ignores the fact that time is a non-renewable asset, and that the timing of events and the repayment of capital is a function of time. As a counter to this argument one can say that it is a moot point whether a Joule of energy today is worth more to you than a Joule in ten years' time. (It is a matter of observation that, because of inflation, the true IRR of industrial economies during the last half century has been between 0 and 2 per cent. Thus we would have been better advised to set our sights on a longer time-horizon.) For example, the reader of this book may be extremely interested in the idea, that by a certain investment today, he can secure for himself a Joule of energy at no cost at all ten years hence, and that that situation would continue for many years. His decision process may be influenced essentially by three factors: firstly how expensive would be the investment today; secondly how long the device would last him and thirdly his mental view of future fossil and fissile fuel prices. Certainly, in a climate of opinion which held that the price of energy from fossil or fissile sources was going to drop and drop as time goes on, one's interest in investment in a solar energy system would undoubtedly diminish. However, there seems to be an almost 100 per cent agreement, that for the mid-term anyway, we are bound to see a rise in prices of energy, and for the long-term little hope of amelioration. Under such circumstances, the attraction of a current investment to be set against the future free income is not inconsiderable.

Perhaps one should conclude this section by pointing out that those who seek to demonstrate the attractiveness or the viability of their bioenergy systems should look at it from both a cost benefit point of view, using the rules of that game, and an energy analysis point of view bearing strictly in mind the rules of that game. Only with the development of data can we test theory against practice. Some impression may be gained of the relation between net energy and cost by looking at Table 6.7, Section 6.5. The higher the net energy (GJ/ha) the lower is the estimated cost, with the point of futility around $10/GJ.

Notwithstanding the foregoing remarks, it will be of considerable interest to look at what sort of economic evaluations have emerged from various studies of bioenergy systems.

6.4 Some conventional economic costings

Traditional economic costing is based on drawing an envelope around the system being studied, and counting inputs and outputs. Since inputs are costed in money terms, they reflect all costs upstream, so no error is compounded from that point of view. However, there is no feed-back from output costs to input costs, so that a changing energy cost environment is not reflected in new cost figures. In systems analysis this is referred to as an open-ended calculation within a limited system boundary. Nevertheless, it is instructive to see what cost figures emerge from calculations of this nature. The following data are all taken from the literature and all too often refer to hypothetical rather than existing systems. In these days of significant inflation, prices should, to be meaningful, pertain to the year of study. In some cases where this information has not been given explicity, we interpreted it as best we could.

The following costings tend to consist of two components: the acquisition of the basic organic raw material, and its subsequent conversion to fuel. The financial cost of procuring the biomass will be dealt with here, and will be seen to vary enormously from zero for wood and dung in rural Africa, India and South-East Asia to prohibitively high values in parts of the West, where the climatic regime is often unfavourable for good yields and where the *per capita* agricultural land area is low.

As in Chapter 3 the biomass can be subdivided into energy crops and organic wastes, and, as might be expected, the production, harvesting, and transportation costs estimated for what are often as yet hypothetical schemes can show marked differences, as exemplified in Tables 6.3 and 6.4.

The wide discrepancies in Table 6.4, particularly in the Californian study, arise principally from the fact that whereas residues in large quan-

Table 6.3 Estimated attainable yields and costs of biomass energy crops

Biomass crop	*Yield (t/ha)*	*Cost/ dry t (US $ 1975/6)*	*Climate*	*Reference*
Sugar cane	*54*	*11.50*	*Tropical*	(21)
Various crops (annual & perennial)	68	14.35–17.65	Sub-Trop.	(47)
Short rotation hardwood	18–22.5	14.55–17.90	Temperate	(147, 148)
Warm season grass	20.5	15.35–18.90	Sub-Trop.	(147, 148)
Elephant grass	68	17.90	Tropical	(130)
Sugar cane	44	18–23	Tropical	(130)
Napier grass	50–60	18.20	Sub-Trop.; Trop.	(149)
Eucalyptus (short rotation)	16	18.90	Temperate	(130)
Short rotation hardwood	11.3–27.2	25.05–40.45	Temp; Sub-Trop.	(48)
Corn	19.3	29.00	Temperate	(149)
Cassava (tubers)	14.5	29.20	Tropical	(21)
Sycamore (short rotation)	16.2	31.00	Temperate	(149)
Cassava (tubers and tops)	29.5	31.80	Tropical	(130)
Microalgae	approx 50	approx 40[a]	Sub-Trop.	(60, 63)
Kenaf	30	41.40	Sub-Trop.	(130)

[a]The microalgae are involved in effluent treatment and so a credit for this may be subtracted from the cost of algae production.

tities can be collected at reasonably low cost, where they are scattered over a large area and are present in small quantities the costs of collection and transportation to an energy conversion site increase the overall costs. In the case of straw especially, its other uses are often competitive, e.g. as a livestock feed ingredient, for cellulose and particle board manufacture etc., thus pushing up the value of the basic raw material itself. Often, however, wastes can almost legitimately be attributed a negative cost where they otherwise have to be treated and disposed of to comply with environmental legislation. Thus in the UK the disposal of treated domestic wastes costs upwards of £6.50 ($13)/t [76], while that for agricultural wastes from intensive livestock farms is approximately £1($2)/t [85]. Where a credit value can be ascribed to a waste product like this, the economics of any bio-

Table 6.4 Estimated collection and transportation costs of organic wastes in various locations

Type of waste	*Cost/dry t (US $)*	*$/GJ*	*Location*	*Reference*
Dung, crop residues, forestry wastes	0–5.50	0–0.30	Rural Africa, India	(90, 150)
Dry mill residues	1.10–121.25	0.05–7.15	California	(151)
Forestry residues	3.15–45.20	0.02–2.65	California	(151)
Straw	7.30–8.40	0.45–0.50	Queensland	(129)
Low moisture field residues	15.15–36.95	0.90–2.15	California	(151)
High moisture field residues	26.30–62.00	1.55–3.65	California	(151)
Orchard prunings	29.95–63.00	1.75–3.70	California	(151)
Straw	approx 30–40	1.85–2.45 approx	Denmark	(152)

conversion process to produce fuel will consequently be that much more favourable, as will be discovered in the following section (6.5).

Returning firstly to the terrestrial biomass plantation concept discussed in Sections 3.1 and 5.2, detailed cost estimates have been performed on both the system proposed by the Stanford Research Institute [46] (see Table 5.2) and that of the Mitre Corporation's silviculture farms [48] (Table 5.3). Much of the work for each of the studies was performed by Inman and Alich who updated the final cost estimate for the original SRI analysis by approximately 50 per cent in a subsequent paper [47]. Therefore, this adjustment has been incorporated in the following costs used in the present book.

The SRI cost breakdown consists of four main features: the annual cost of labour and of fuels; of chemicals; and of farm machinery. Labour and fuel costs combined include planting and nurturing of the crop, harvesting, and hauling. This last requisite of transportation to the plantation gate accounts for roughly one-third of the total labour and fuel costs; with irrigation taking up 12 per cent, over 50 per cent of which is for pumping costs. At about $5.25/hr labour costs, and diesel fuel at 12.3¢/litre and electricity at 1.5¢/kWh (all updated values), fuel costs amount to 28 per cent labour costs overall. With respect to farm chemicals, anhydrous ammonia, phosphorus (P_2O_5), potassium (K_2O), herbicides, insecticides and fungicides are cited, with 60 per cent of the total cost being apportioned to the ammonia because

Table 6.5 Annual biomass cost summary for hypothetical plantation (after [46, 47])

Item	*Cost/ha. yr, 1975* $	*% total*
Labour and fuel	218.10	20
Farm chemicals	403.46	37
Farm machinery	70.65	6
Irrigation equipment (installed)	112.50	10
Land charges	131.25	12
Water charges (7 ha feet at $18.75 per ha foot)	131.25	12
Seed	15.00	1
Interest (7½%) on water and seed	10.97	1
Total costs, per hectare year	1093.18	
Total cost, per t dry biomass (at 68 dry t per ha year)	16.08 = $ 0.93/GJ	

of the large quantity of nitrogenous fertilizer necessary to fulfil yield expectations. Farm machinery expenditure, excluding irrigation equipment, is calculated on an annual basis taking into account depreciation over the equipments' amortization periods, maintenance repairs at 15 per cent of the annual cost of depreciation and interest (7¾ per cent pa), and property taxes. Machines used in the transportation of the biomass to the plant gate again incur the greatest costs from this section. A summary table of updated costs and various percentages involved is given in Table 6.5. Unfortunately labour and fuel costs are not separately identified.

Labour and fuel costs amount to 20 per cent of the total, farm chemicals 37 per cent, irrigation equipment 10 per cent, land and water charges 25 per cent, and farm machinery only 6 per cent.

The above values can be roughly compared with a more disaggregated breakdown for two of the favourable US sites for silvicultural plantations (Table 6.6). Exact comparisons cannot be made because the analysis is not broken down in an identical manner to that of the SRI study.

The Mitre Corporation authors believe that there will be future decreasing production costs principally due to increased productivity levels developed through an organized field research programme. Capital costs comprise only 9–11 per cent of total costs, irrigation and fertilization up to 40 per cent, transportation 10 per cent, land costs 5–7 per cent. There are

Table 6.6 Breakdown of total production costs at Wisconsin and Louisiana sites, current and future cost and technology scenarios for silvicultural plantations (after [48])

	Per cent of total costs			
	Wisconsin		*Louisiana*	
Cost category	*Current*	*Future*	*Current*	*Future*
Planning, supervision field support	5.7	8.1	8.5	9.6
Land lease	6.9	4.9	5.2	3.7
Land clearing/ preparation	4.2	3.0	2.9	2.0
Roads	0.5	0.5	0.3	0.3
Planting	2.0	1.1	2.6	1.2
Irrigation	19.9	14.1	12.3	8.3
Fertilization	16.5	25.0	26.8	37.1
Weed control	0.6	0.4	0.5	0.3
Harvesting	4.6	5.0	4.6	5.5
Green haul/ storage	2.3	4.0	2.5	3.0
Loading	1.5	2.3	1.9	2.3
Transportation	10.3	12.2	9.3	9.2
Interest on debt	3.8	2.7	3.4	2.2
Taxes	11.4	9.2	10.3	7.9
Return to investors	10.7	8.6	9.7	7.5
Salvage value	(0.5)	(0.4)	(0.4)	(0.4)
Total cost ($) per t dry biomass	36.85	24.45	25.05	20.70
and per GJ	2.10	1.40	1.45	1.20

thus some differences between Tables 6.5 and 6.6, but apparently not so significant as to render them incompatible. Finally, they find that costs at highly productive sites are generally most sensitive to productivity-independent cost items such as fertilization, while costs at less productive localities are by contrast more influenced by items related to the level of productivity, e.g. irrigation and land costs (see Table 6.6 for the highly productive Louisiana and less productive Wisconsin sites) [48]. Thus the cost of energy from these intensive systems is envisaged to be about $1–2/GJ of heat.

6.5 Costs of biomass fuels

6.5.1 'Biofuels' in general

The cost of biomass acquisition is, of course, only the beginning of the total cost of producing photobiological fuels, and, before quoting more itemized costings of specific processes it might be useful to return to the Australian study outlined in Table 5.12. The economic, rather than the energy, costs envisaged for these seven processes are presented in Table 6.7, along with the cost of more conventional fossil fuels for comparison in the particular situation of Australia.

The cost of those fuels where by-products are also produced includes a credit for the by-products. Table 6.7 indicates that on a $ per GJ basis liquid and gaseous photobiological fuels would be even more expensive than synthetic fuels derived from coal, and more so when contrasted with crude oil and natural gas at 1975 prices. A closer appraisal of such possibilities will be attempted in Section 6.6.

6.5.2 Ethanol

A further examination of the ambitious Brazilian national alcohol programme is merited here as probably being the pioneering large scale attempt at a feasible, practical, fuels from biomass scheme. During the past 10 years the Brazilians' demand for oil has risen 10.6 per cent annually, 82 per cent of which has to be imported (along with 67 per cent of their coal). Naturally, this has done nothing to help the country's balance of payments problem and as mentioned in Section 3.1, the Government has initiated their alcohol production programme (Proalcool) from the fermentation of indigenous agricultural crops, and additionally has sponsored research into the potential for ethanol, and, to a lesser extent, methanol utilization in gas turbines and Otto and Diesel engines. Studies carried out by the Brazilian Centro de Tecnologia Promon (CTP) have shown that both the cassava and sugar cane plantation/distillery systems are clear net energy winners, approximately to the same extent where both sets of conditions are optimized. Either sugar cane bagasse or firewood are the principal energy suppliers, while the electricity requirements are met by on-site generation via steam turbines or, alternatively, by acceptance from the grid. The steam to operate the turbines is raised by the combustion of bagasse or firewood, while 90 per cent of Brazil's 25 GW installed capacity accrues from hydroelectric plant, thus lowering the GER for electricity substantially, and improving the overall net energy analyses. What this means in terms of economics is shown in Tables 6.8 and 6.9.

Table 6.7 Economic costs of photobiological fuels and prices of fossil fuels (1975 figs) [128]

Fuel	*Raw material*	*Process*	*Saleable by-products*		*Cost ($/t)*	*Cost ($/GJ)*	*Net energy (GJ/ha)*
			As fuel	*Other*			
Alcohol	Cassava tops & tubers	Enzyme hydrosis, Batch fermentation	—	Fibre (animal feed); fusel oils	250	8.4	+80
Alcohol	*Eucalyptus*	Acid Hydrolysis, Batch fermentation	—	—	400	13.4	−452
Alcohol	*Eucalyptus*	Enzyme Hydrolysis, Batch fermentation	—	—	600	20.1	−452
Methane	Cereal straw	Anaerobic digestion	—	Biomass slurry	235	4.2	+7
Methane	*Eucalyptus*	Anaerobic digestion	—	Biomass slurry	310	5.5	+84
Pyrolytic oil	Cereal straw	Flash pyrolysis	Char	—	75	3.3	+11
Pyrolytic oil	*Eucalyptus*	Flash pyrolysis	Char	—	100	4.3	+131
Alcohol		Non-biological Chemical synthesis	—	—	275	9.3	?

Fossil fuels (1975 costs)	Cost ($/GJ)		
Kuwait crude oil	1.25	Diesel Fuel (untaxed)	2.0
Syncrude from coal	1.2–1.9	No. 6 Fuel oil	1.7
Petrol (taxed)	4.45	Natural Gas	1.15

Table 6.8 Economic breakdown of anhydrous alcohol production from sugar cane juice in Brazil for 21000 t per yr plant [21] (1975 US $ 10^6 assumed)

Investment	*US $10^6*	
Fixed investment	13.08	
Working capital	1.54	
Costs (US $/t of alcohol)	*$*	%
Sugar cane raw material (at $11.54/t)	219	44
Chemicals	6	1
Utilities–water	1	0.2
Credit for by-products	–(28)	–(5.5)
Labour	14	3
Maintenance	14	3
Depreciation	40	8
Income tax	32	6
Value added tax, other taxes, insurance, administrative expenses	125	25
Return on investment (12%/yr, DCF)	78	15.5
Total product cost	501	100.1

The above result means a cost of 0.5$/kg alcohol or 16.8$/GJ. This is, incidentally, a higher price than the most heavily taxed petrol in Europe. Tables 6.8 and 6.9 refer to a 150 m^3/d (or 117 t/d) distillery in which by-product credit represents the difference between the cost of direct application of stillage as fertilizer and the sales of hydrated ethanol and fusel oil which are also formed in the processes. In both cases the operational lifetime of the distillery is taken as 15 years, but whereas the cassava-based operation functions for 330 days/year, that founded on sugar cane operates for only 180 days/year (49500 m^3/year absolute ethanol as against 27 000 m^3/year). This results in the necessity for a larger proportion of working capital in the case of the latter due to the requirement for an ethanol storage capacity adequate for an even continuous year-round supply, including the off-season sugar cane growth period. The higher capital charges in cassava distilleries arise initially from investment in additional equipment for starch conversion to sugar and for electric power generation, while in all systems the cost contribution of the raw material feedstock ranges from 42–48 per cent, emphasizing the importance of achieving sustained high-level plantation yields.

Table 6.9 Economic breakdown of anhydrous alcohol production from cassava for 386 000 t per yr plant [21]

	Base distillery		*Continuous conversion*	*Cassava stalks as fuel*		
				Mechanical drying	*Sun drying*	*Methane from stillage as fuel*
INVESTMENT						
Fixed capital	15.75		14.25	21.15	15.75	17.92
Working capital	1.06		1.04	1.12	1.06	1.08
COST (as $/t of alcohol)	$	%	$	$	$	$
Raw materials:						
Cassava roots at $29.20/t	256	47	256	256	256	256
Enzymes and other chemicals	38	7	38	38	38	38
Utilities:						
Water	0.5	0.1	0.4	0.5	0.5	0.5
Power at $28.50/MWh	16.5	3	14	16.5	16.5	17
Firewood at $7.00/t	15	3	13	23 (b)	10 (c)	13
By-products credit	−(23.5)	−(4)	−(23)	−(25.5)	−(23)	−(24)
Labour	12	2	12	12	12	12
Maintenance	14	3	12.5	18.5	14	16
Depreciation	41	7	35	52.5	40	44
Income tax	16	3	15	21.5	16	18
Other taxes, insurance and administrative expenses	127.5	23	123.5	142	125	132
Return on investment (for 12%/yr DCF)	37	7	35	50	38	42.5
Total product cost (a)	550.0		531.4	605.0	544.0	565.0

(a) The confidence level of cost figures for the alternative cases shown, is less than for the base-distillery, due to insufficient basic design data.
(b) Cassava stalks at transportation cost: $3.10 pet t.
(c) Cassava stalks cost is the sum of costs of transportation and labour for sun-drying: $3.50 per t.

The route to ethanol via cassava has thus been costed for a number of feasible modified scenarios (Table 6.9) involving the continuous conversion of the starchy raw materials, the use of dried cassava stalks as fuel, and the production of methane via anaerobic digestion of the cassava stillage rather than utilizing the stillage as a fertilizer, as indicated.

The respective costs of sugar cane alcohol and cassava alcohol are thus \$501/t or \$397/m^3 (\$16.8/GJ) and \$550/t or \$249/m^3 (\$18.1/GJ) in normal practice. These very high costs are acceptable only because of the high selling price of petrol in Brazil at \$470/m^3 (\$19.9/GJ). From these figures the impact which fermentation alcohol can make on the country's economy will largely be determined by political decisions. For example, if the selling price of alcohol currently (1977) at \$339/m^3 (\$14.3/GJ) as fixed by the Government, were set to the real cost and the Government absorbed some of the ensuing blending and distribution costs, a profit margin could well be attainable by ethanol manufacturers, fuel distributors, and retailers. The crucial points here are how much the price of petroleum will increase in real terms over the next few years to enable fermentation alcohol to become economically more competitive, and indeed to what extent technological improvements will lower production costs, and finally what is the optimum plant combination of sugar cane and cassava processing to arrive at a minimum alcohol production cost figure.

6.5.3 Methane

Having thus dealt with the economics of fuel production based on the conversion of photosynthates formed as a result of energy crop cultivation, it must be remembered that in many nations of both the developed and developing world it is still more sensible to investigate 'fuels from waste' programmes. The concept of what is a 'waste' in turn varies from country to country and culture to culture, so than animal dung, for example, may be classed as a waste in the West, but is an important source of energy and fertilizer in those numerous countries which produce and consume negligible quantities of coal, oil, gas, or electricity.

Table 4.1 of Section 4.2 presented the basic data pertinent to the Indian Gobar 'biogas' generator utilizing livestock waste in that vast subcontinent. Upwards of 75 000 plants are now operating, some more successfully than others, in rural India at an animal *per capita* figure of approximately 0.5. Although most of the current 'biogas' generators in use are strictly family-size, their cost of US \$200–250 and input needs of from three to five cattle to meet normal cooking and lighting needs puts them beyond the reach of most rural Indian families. In effect the domestic energy

Table 6.10 Economics of a village level plant of 170 m^3 gas/day capacity [90]

Costs	*US $*
A. Investment	
a. Plant of 170 m^3/day capacity	4000
b. Plant-site kitchens	500
c. Plant-site washing facilities	200
d. Compressor	300
e. 250 Cylinders	2500
f. 100 Burners	1000
g. Delivery carts	500
h. Land and preparation for dying beds	1000
Total investment costs	10000
B. Operating costs per annum	
a. Purchase of dung – 500 kg at $6.0/t day (6.0 × 365)/2 =	1100
b. Plant and compressor maintenance	400
c. Staff	
i. Manager/accountant ii. Dung collection and feeding iii. Gas distribution iv. Water procurement v. Kitchen maintenance	1000
d. Total operating costs	2500
C. Receipts per annum	
a. Sale of gas	
49 500 m^3 of bottled gas at $4.60/100 m^3	2275
12 600 m^3 of gas at $3.5/100 m^3	440
b. Sale of fertilizer: 263 t of N at $450/t	1180
Total sales	3895
D. Gross annual earnings (receipts – operating costs)	1395

needs of some 90 million people (14 per cent of rural householders) could be met at an investment of $3 billion in this way, but this would satisfy less than 20 per cent of total rural domestic energy demand. S. and K. Parikh made a cost-benefit analysis of 'biogas' generation at the village level envisaging that waste from even the smallest small-holding could be used in the corporate whole, any excess gas produced in the summer superfluous to one family unit to be utilized for the benefit of others. They assume the usual economy of scale benefits of land sequestration, capital investment, and labour requirement would apply, with the proviso that participation would be unanimous because of the benefits being to, and accepted by, all strata of the community. They view this principally through the introduction of a pricing policy whereby the prices of the process inputs and outputs are fixed in relation to other sources of these materials. For example, they argue that the unit price of gas should not be greater than the price of the same quantity of energy obtained from other competing fuel sources, and the price of by-product fertilizer must not exceed that of equivalent nitrogen derived from chemical sources.

Given such a pricing policy the costs and benefits of a village-scale plant of 170m^3 'biogas'/day capacity are shown in Table 6.10. From this it can been seen that, even at an interest rate of 12 per cent, the amortized capital investment over the approximate 15 year life-span of the equipment is more than satisfied by the annual profits.

Projecting rather tentatively ahead to the year 2000 they postulate that 1.6 million such plants could meet 90 per cent of rural domestic energy demand at a total investment cost of $16 billion, a figure which includes the necessary ancillary equipment cost of burners, plant-site kitchens, etc. This compares with an investment cost for coal-mine and railway network development of around $9.5 billion for the sequestration and distribution of the 230 million t of coal required for the provision of an equivalent quantity of rural domestic energy throughout the subcontinent. Since the annual operating costs of coalmining and transportation will be significantly greater than the comparatively negligible cost of 'biogas' plant operation and maintenance, the difference in capital cost will quite rapidly be closed and, presuming coal prices rise in real terms (as they almost certainly will in India), in the medium-long term the overall cost of coal energy will be substantially higher than that of 'biogas'. Therefore the desirability of implementing a co-ordinated 'biogas'-based energy strategy for rural India as soon as possible would seem to be unquestioned by the application of any criteria, whether economic, energetic, environmental or social; and it is to be hoped that such a scheme will be instigated to help alleviate the plight of the millions of that country's population currently living in conditions of quite abject poverty and destitution. Most of the alternatives would appear to offer little hope of redressing the existing imbalance, not so much between the rich and the poor, but between the poor and the absolutely destitute.

As has been observed previously in Sections 4.2 and 5.4 the anaerobic digestion of waste organic matter shows much potential for energy recovery also in large urban communities. Table 6.11 presents the relevant economic data for methane production from municipal solid waste according to the US Dynatech R/D Company's computer model optimization for minimization of the unit cost of gas production.

Interestingly, the minimization of the unit cost of gas production was selected as the optimization criterion due to the dearth of absolute information on the scale factors relating to the sale price of gas, other

Table 6.11 Summary of Dynatech's base-line gas cost items (1974/5 figs) [99]

A. Contribution of capital costs to gas cost	$
Preparation	0.46
Digestion	0.54
Dewatering	0.34
Gas scrubbing	0.35
Other	0.29
	1.98/GJ (gross)
B. Contribution of operating costs to gas cost	
Power ($0.015/kWh)	0.20
Heating ($1.9/GJ)	0.28
Cooling water ($0.005/m^3)	0.01
Chemicals	0.01
Labour	0.39
Administrative and general overheads	0.24
Supplies	0.27
Local taxes and insurance	0.38
	1.78
C. Penalties	
Filter cake ($33.10/t)	1.43
Waste water from filter ($0.26/m^3)	0.03
Waste rejected by Trommel screen and air classifier ($3.85/t)	0.25
	1.71
D. Credits	
Fresh waste ($11.75/t)	2.67
Sewage sludge ($55.15/t)	0.47
Scrap iron from resource recovery ($27.95/t)	0.44
	3.58/GJ (gross) gaseous product
Gas cost/GJ (gross) = A + B + C − D = $1.89	

credits, or penalties. Were these prices to be satisfactorily fixed then the more usual requirement, that of maximization of profits, would closely approximate the results for minimization of gas costs in any event. The exercise was carried out using a system based on 'most likely' performance and costing information, to arrive at a baseline calculation where the total capital for constructing a plant dealing with an average of 1000 t municipal waste/day is about $21.8 million – to give a final gas cost of $1.89/GJ (gross). This figure would certainly be below a reasonable selling price and ensure a financial profit for the scheme, though probably not a large one, with current (early 1978) US selling prices to the consumer of around $2/GJ and a UK figure of about $2.50/GJ.

It would not be prudent to terminate an economic appraisal of energy recovery from waste processes without reference being made to the excellent in depth analysis carried out by Porteous of the Open University, UK, on a variety of incineration, pyrolysis, and anaerobic digestion processes. He concludes each analysis with a disposal cost/t of input domestic refuse allowing for any revenue accruing from sales of heat, gas, electrical power and materials recovery. These costs on a mid-1975 basis are shown in Table 6.12.

From the evaluations made in Table 6.12 it is concluded that energy recuperation is not particularly attractive from an economic standpoint at

Table 6.12 Refuse disposal cost following various treatments (1975 £)

Process	*£/t input refuse* [76]
1. 1333 t/day non-recuperative incineration (continuous operation)	6.50
2. 1333 t/day recuperative incineration (selling heat at the boiler stop valve; 7-day week operation)	6.49–8.65
3. 1333 t/day recuperative incineration (selling electricity (30 MW continuously available) at base price of 0.6p/kWh; 7-day week operation)	7.322
4. 1333 t/day recuperative incineration (selling electricity (20 MW continuously available) at base price of 0.6p/kWh; 7-day week operation)	8.58
5. 250 t/day pyrolysis plant producing 3.165 GJ/t input refuse	6.115
6. Pyrolysis and steam raising plant (900 t/day maximum capacity; 770 t/day minimum capacity)	6.365

present, particularly where electricity is produced since, in the UK at least, there is already an over-abundance of installed power capacity. However, circumstances will obviously be different in other locations. One possibility which presents itself for more detailed examination is that of utilizing heat recovery techniques from municipal refuse within district heating schemes in combination with coal. This may be an attractive proposition in high density areas within the built environment, while the supplementary use of refuse in coal-fired boilers, especially for industrial steam raising, also has its merits, since the accompanying coal would act as a buffer to reduce the emission of corrosive gases and assist in raising the thermal efficiency of direct refuse firing alone. As with many of the innovative system approaches their values are often site-specific, and the expounding of generalizations becomes an exercise in futility. While energy recovery from urban refuse is a desirable and laudable objective, its implementation on any large scale depends upon several energetic, economic, and environmental criteria. The fact that in many cases it does not do so now by no means implies that it will not do so ten, five, or even three years into the future!

6.5.4 Electricity

Electricity may not be the most appropriate product of biomass conversion processes. But the direct combustion of wood for electrical power generation is the most advanced technology at present available for its utilization as an energy source, and even though it might be currently non-competitive in many areas in its own right, it could be so in combination with coal in significant proportions to lower sulphurous emissions and meet increasingly demanding pollution standards.

What might almost be described as the definitive study on the prospects for silviculture biomass plantations and their energy conversion products, at least within the United States, has been conducted by the Mitre Corporation. Their economic appraisal demonstrates just how dependent on location the possibilities for a particular scheme are. In New England (situated in the north-east), for example, the wood feedstock obtainable from standing forests for 220 MW plants is almost competitive with present-day 1 GW coal-fired power plants. Smaller plants (just over 50 MW capacity) already exist in some of the northern New England states, to the detriment of coal, owing to the latter's comparatively high current cost. The retro-fitting of small capacity oil/gas-fired power plants wholly or partially to combust wood would be economically viable as against using coal up to around 150 MW output. A 55 MW fired plant would annually require ¼ million t wood on a dry weight basis, while a 220 MW plant's demand would be four times as great, sizes larger than this being unfeasible due to biomass supply limitations and diseconomies of scale [52].

This kind of economic approach leads into Section 6.6 where present and future price comparisons are made; not without some trepidation, one hastens to add.

6.6 Relative prices of biological and other fuels

Before embarking on a general appraisal of the economics of the conventional and not so conventional energy supply options, with particular relevance to those from biological sources, it is interesting to look at the Mitre Corporation's projections. Their main philosophy is that wood-derived fuel prices are expected to decrease in real terms due to improved technologies increasing biomass productivity (and thus lowering feedstock costs), while projected prices for conventional fuel feedstocks are assumed to rise at rates above that of the mean inflation rate. Given these

Table 6.13 Future competitiveness of energy derived from wood [48, 52]

Product	*Year in which wood becomes price-competitive*		*More conventional competitive sources*	
	Small plant	*Large plant*	*Current*	*Future*
Ammonia	Now	Now	Natural gas	Coal
Methanol			Natural gas	Coal
(Los Angeles Market)	1983	Now		
(Gulf Coast Market)	1983	Now		
Ethanol	1992 app	1985 app	Ethylene	Ethylene
Medium-energy fuel gas	1998+	1989–1998	—	Coal
Substitute natural gas	21st Century	1993–2001	—	Coal
Electricity			Coal/Nuclear	Coal/Nuclear
(small wood-fired 55 MW–220 MW) versus (large coal-fired (1GW) plants)	21st Century if at all			

assumptions, the findings are inevitable. However, as indicated in Section 6.1, we have some reservations about such an interpretation of events. The results are summarized in Table 6.13. Economies of scale are assumed. Two scenarios for biomass costing are employed to provide an 'optimistic' (low) price projection and a 'pessimistic' (high) projection.

The table is naturally US-based and should be treated with this limitation in mind, but it is the result of very carefully calculated figures. Nevertheless, it is heavily predicated on their underlying philosophy. The results suggest than ethanol will be an economically competitive biomass product before, for example, SNG.

It would be true to say that speculation about future energy prices in great detail serves little purpose, particularly so when there is still so much controversy about current energy costs from different sources; the difference between costs and prices being the difference between the monetary value requirement to produce a given quantity and quality of energy and that required of the consumer to obtain it. One must also be careful to compare 'like' with 'like', as we showed earlier (see Table 6.2). This area becomes particularly dangerous (and sometimes explosive) when evaluating the cost of a nuclear power system, the capital costs for which tend to exclude those for ancillary waste disposal facilities and fuel reprocessing plant. It is not proposed to enter the great nuclear debate here but the capital costs for their construction appear to be escalating to such an extent that no one seems to actually know the absolute figures, but suffice it to put a value somewhat in excess of £650 ($1300)/kW installed capacity, without considering capital for safeguards, reprocessing and storage.

In the developed nations of the world (at least in the more enlightened of them) research and development is increasingly being extended to unconventional energy supplies other than nuclear, from what might be termed ambient energy sources. Many of these are geared principally for electricity generation and current estimated costs in US $ per installed kilowatt are estimated as follows: [150, 153–155]

Wave: 600–1600 | Wind: 300–1200
Tidal: 600 upwards | Ocean Thermal (OTEC): 1500–2500 approx.
Centralized solar photovoltaics: 600–950
Biomass plantations: 550

It must be stressed that these evaluations can be no more than reasonable approximations but it would appear than biomass combustion followed by electricity generation does not suffer by comparison with other options, and is more economically attractive than most. However these costs are sensitive to many parameters and situations which are operating today and which will continue to operate in the future. Furthermore, it is not always obvious from the sources from which costing information is obtained exactly what particular concept is being employed to arrive at the final figures, although unless otherwise intimated the figures are assumed to be average and not marginal costs. Finally, to attempt to bring some semblance of order out of all this apparent confusion it may be worthwhile quoting the US Department of Energy's estimates of the factors by which some alternative energy sources would provide electricity in excess of current conventional electricity prices. These are as follows relative to an electricity price of unity: wind power systems 1–2, biomass 2–4, OTEC 4–5, solar power towers (heliostat systems) 5–10, photovoltaics 20–40 [156]. Not everyone would agree with these figures, particularly for photovoltaics (the costs of which seem to be decreasing rapidly), but they do give some idea of the present orders of magnitude involved. A somewhat different situation occurs in many regions within the under-developed world where on-site solar power systems are economically competitive at present, especially of course in unelectrified rural areas. Even where electrification has taken place the economics may be prohibitive. In the USA for instance the cost of electricity to the consumer ranges from 3–10 cents per kWh, while in Third World cities it can be as high as 45 cents per kWh. The situation is exacerbated in the rural Third World where diesel generator sets often provide the only available source at some $1 +/kWh or batteries at some $12/kWh. The economic prospects for solar applications are therefore excellent under these conditions when solar-thermal power systems already have the capacity to deliver electricity at under $4/kWh now, and these and/or photovoltaic systems could well be capable of electricity generation at well under $1/kWh within a few years by virtue of the rapidly advancing technologies and mass production opportunities involved. An excellent report on the difficulties in applying solar energy technologies within developing countries prepared for the Dutch Ministry for Development Cooperation includes costings for both conventional and nonconventional energy sources within a selected country from each of Latin America (Columbia), Africa (Upper Volta), and Asia (India) [150]. Prices are given for electricity from the grids, engine generator sets and batteries, mechanical power from combustion engines, draught animals, and human labour, and thermal energy derived from coal, oil and butane, as well as for thermal solar energy systems. The price ranges, highly dependent on location and circumstances, are such that the actual figures serve little purpose in being quoted herein, but the conclusions drawn concerning the economic feasibility of various functions performed via solar energy applications are interesting. Solar water pumping, solar powered cold stores, solar water distillation and disinfection, recharging batteries using solar cells, and providing solar electricity in the absence of a grid system are all cost-effective, while solar crop drying, solar cooking (as against utilizing wood), and solar electricity in competition with grid supply are not. Thus the potential for such applications within the Third World is by no means to be dismissed, and the enormous impact which could be made more specifically in rural India via the anaerobic digestion of dung has already been demonstrated in the previous section.

Returning to Western society and to applications of biological energy generating systems, a number of schemes were referred to in Sections 6.4 and 6.5. Where these are related to waste treatment, as in the anaerobic digestion or recuperative incineration of municipal wastes, energy supply is not the sole criterion, and therefore the cost competitiveness of the resulting fuel against conventional sources may not be a crucial factor. However, if industrial ethanol, for example, were to be fermented from grain or cellulosic raw materials then it would certainly need to be as economic as the currently used chemical synthesis via ethylene, which in turn is derived from petroleum and natural gas. The production cost is obviously closely related to the raw material cost and, in the United States, the price of ethylene more than doubled in the two years from 1972 to 1974 after remaining stable for a considerable length of time previously. Since the real price of petroleum-based raw materials is envisaged to rise then the price of ethylene will rise in concert, as will the price of ethanol. In 1975 Miller [10] considered that the price of grain would increase at a rate less than that of ethylene, so that within a decade fermentation ethanol should be less expensive to produce than ethanol derived from fossil fuels.

The overall capital and processing costs for utilizing wood as a raw material rather than grain are estimated to incur an extra 10 per cent financial outlay, and so the grain-ethanol route would be likely to become cost-effective at an earlier date. Table 6.12 indicated that ethanol from wood would be price-competitive for a large plant by about 1985, and for a small plant roughly seven years later. From this it may indeed be deduced that ethanol from grain could be price-competitive under the most favourable conditions sometime in the early 1980's. We should stress that all these forecasts call for exogenous assumptions.

One of the hardest things to forecast is future fuel prices, or even relative prices, but factors affecting costs may be worth examining. In the case of coal, the most abundant of our fossil fuel reserves, its costs are closely tied for a long time to come to surface mining costs in the USA and elsewhere. Manual labour costs in such a situation will probably rise in real terms and technological developments may not be sufficient to prevent the real price of coal from increasing in most countries, particularly in the West. With respect to petroleum some nations are undoubtedly better placed than others, but oil will increasingly be treasured, but for the hydrocarbon molecule rather than its energy content. The price of crude oil will rise, at a rate considerably higher than that of coal. A similar situation exists for natural gas as for petroleum, future shortfalls being made good by the manufacture of substitute natural gas (SNG) from coal and maybe coal and nuclear energy. When SNG does become the marginal supply then the price of gas will almost certainly have increased by over 100 per cent from that of today even in real terms. It therefore would seem not unreasonable to suggest that as a general rule energy prices from the more conventional sources will rise in real terms until 2000 or thereabouts, and, depending on the state of the nuclear and solar industries in particular, could stabilize somewhat thereafter. With each succeeding price increase of conventional energy the competitive edge of the alternative energy sources, including biomass energy, will be sharpened. Yet as we have pointed out these systems will initially be costed through, not against conventional energy costs. Hence if biomass systems are to succeed, they must continually improve technologically in net energy output per hectare.

In the Third World of course, where the opportunities are that much greater, for solar exploitation in particular, the impact is liable to be quite considerable. The poorer nations of the world will decreasingly be able to afford even modest quantities of imported petroleum, civilian nuclear electricity would seem to be grossly inappropriate. Thus, if living standards in these countries are to improve at all then the energy necessary to achieve such aspirations must be available in relatively consistent quantities and at a reasonable price. Ambient and biological energy sources would therefore appear to be the only sensible solutions in these circumstances, and, even if they are less likely to make an outstanding contribution in the West, then Western industrialists should still bear in mind the export potential which is undoubtedly available in this expanding area, as many indeed are already well aware.

Of these innovations the integrated systems either allied to waste treatment or food and fibre production show particular promise. The Brazilian national alcohol programme is in many ways a pioneering one, and should it succeed overwhelmingly then other governments might well be prepared to embark on similar, though perhaps less ambitious, schemes in its wake. In the final analysis, the economic feasibility of projects such as this can only be proved or disproved under field conditions, and so it is to be earnestly hoped that the Brazilian venture does indeed succeed, and is seen to succeed, not only for the benefit of the Brazilians themselves, but for the repercussions which would arise, hopefully leading to the benefit of mankind as a whole.

Chapter 7

Present developments and future prospects

7.1 The state of the art – An overview

Although the vast majority of fuels-from-biomass enterprises are as yet only at the hypothetical or research laboratory stages, the number of processes now in pilot plant, demonstration, and even commercial operation are well into three figures and are increasing rapidly. Many of the latter are related to waste disposal, and are usually run as anti-pollution measures and/or as means of minimizing energy losses. Some have already been mentioned earlier in the text, but a general appraisal of the 'state of the art' might prove useful and is presented here. Areas of key research which are currently under investigation and which are deemed to hold great promise for the future are discussed in Sections 7.2–7.7.

The greatest advances have, not surprisingly, taken place in the United States with its vast natural, financial, educational, and commercial resources, allied to a wide range of differing regional climatic conditions. This accounts for the emphasis on what might be termed Western technology throughout this book, while paradoxically the greatest impact could well be made via adapted low technology scenarios operating in the developing nations of the world.

In the USA, research and development funds evolve from two primary sources, namely, the Environmental Protection Agency (EPA), and what was formerly the Energy Research and Development Administration (ERDA) but is now integrated within the US Department of Energy. EPA has been funding energy development for 11 years and concentrates on four basic areas of endeavour: environmental, technical and economic assessment projects; utilization of municipal wastes as supplementary fuels with coal, oil, and gas; use of municipal wastes as supplementary fuels with sewage sludge; and a study of pyrolysis and purely bioconversion processes. Little effort is being expended on the transformation of naturally occurring biomass since EPA's resources are directed towards the massive 4.5×10^9 t of municipal wastes generated annually in the USA. It is possible that as much as 15 per cent of the country's energy demands could be derived from this waste, but a more realistic target would be in the region of 3 per cent when inefficiencies in waste collection are fully considered [157]. The Department of Energy, on the other hand, has a strong bias towards natural biomass conversion and will shortly be embarking on its final year of a five year fuels-from-biomass energy programme in which the initial goal is the development of appropriate technologies for producing and converting biomass and ultimately to launch an active demonstration programme. Again, four general fields of study are currently being funded: fuels from agricultural and forestry residues, terrestrial biomass production and conversion, marine biomass production and conversion, and more advanced research. It is anticipated that by 1982 final decisions on which technologies are suitable for development and demonstration should have been made.

However, it is particularly within the Third World nations that the energy, and also food crises are most chronically felt. Where land is unsuitable for food production in these regions it has been suggested that plants could be cultivated solely for their energy content [158]. Co-operation between countries would seem to be crucial here. Energy farming is already under investigation in the United States, Canada, Brazil, Australia, Ireland, Israel, Sweden, Norway, France and West Germany, and some of these countries have in fact initiated programmes abroad in India, Mexico, Peru, Thailand, and South Africa. To quote a specific example, Professor William Oswald at the University of California, Berkeley has been invited to participate in special studies of microalgal applications (Section 3.1) within projects envisaged to be mounted in Tunisia, Israel, India, Pakistan, the Philippines, and Jamaica – and this is just the beginning [60]!

To date, European efforts in biological energy production have been more modest than those of the United States, though the Swedes in particular have maintained a great interest in the potential afforded by silvicultural energy exploitation [159]. However, the member nations of the European Economic Community (EEC) in 1975 embarked on a four year investigation into the possibilities for solar energy in general starting with a financial investment of $23 million, of which 6 per cent was to be devoted to a biomass programme and 12 per cent to studies in photobiology, photochemistry, and photoelectrochemistry. However, fully 43 per cent was set aside for research and development on photovoltaic technology. Initial feasibility studies embraced the evaluation of locations, wastes utilization, energy crop cultivation, conversion systems, and subsequent energy and economic analyses. Particular emphasis has been placed on short rotation forestry, forestry wastes, and straw in the climatic constraints of Western Europe. Short rotation forestry is considered to be especially suitable for Ireland. Fourteen trees were evaluated under Irish conditions, including *Pinus* sp, *Eucalyptus* sp, alder, birch, ash, and chestnut, with *Salix* sp, and the poplar being selected as the most likely candidates for energy crop growth [160].

Much of the cereal straw produced in Western European agricultural practice is currently wasted, and so ways of utilizing this potentially valuable resource are being examined in Denmark, France, and West Germany. At present there are 6000–7000 straw furnaces in operation in Denmark, but since two million tonnes of straw remain unused each year there is a great deal of further scope for Danish farmers to heat their houses, while the amount of waste straw arising annually in West Germany is sufficient not only to provide heat for 150 000 farmhouses but also to dry 50 per cent of the yearly grain harvest. The French straw programme is subdivided into geographical regions, and within each region are assessed the quantities produced, and effects on soil fertility of their removal, the economics and energetics involved in harvesting, and the potential for 'biogas' generation.

Elsewhere in the world the ambitious Brazilian national alcohol programme has already been described (Section 3.1) along with the feasibility of energy plantations, terrestrial, marine and freshwater, throughout the earth's sub-tropical and tropical regions. The proliferation of small-scale 'biogas' generators within the Indian subcontinent has been featured in Section 3.2 and these similarly occur in large numbers throughout China. In fact the Sudanese Government (with the assistance of the US National Aeronautics and Space Administration) is also currently experimenting with small-scale digesters for the processing of the thousands of tonnes of water hyacinths mechanically harvested from the White Nile [72]. All the above-mentioned schemes merely form the tip of what is hopefully a potentially very profitable iceberg in terms of energy returns on the initial outlay of research and development effort, manual labour, financial investment, land and natural resource allocation, capital equipment, and of course energy input.

Nevertheless, it is in North America and in the United States in particular that the greatest advances are being made throughout a whole range of technologies. Canada though is also very well placed in that it has a large surplus of timber by virtue of its extensive forests, and the Canadians are looking seriously at the prospects of utilizing this resource, along with straw and municipal wastes, for conversion to methanol. The methanol formed would then be blended at the 15 per cent level with conventional petroleum as a transport fuel in an analogous manner to the Brazilians' plans for ethanol [161]. Methanol can in fact be blended with petroleum at the 25 per cent level, and only tuning adjustments to, for example, an automobile engine need be made to accommodate the blend. Fuel volume consumption would be higher as a result of the lower energy content of methanol, but offsetting this minor inconvenience are the low emission levels, improved combustion efficiency, and increased octane rating. Furthermore, existing engines can even be converted to using pure methanol by decreasing the ratio of air to fuel consumed, from about 14 for gasoline to six for methanol, by recycling more heat from the exhaust carburettor and providing for cold starts – and all for a little over $100 [7].

Bearing in mind that an acute shortage of oil from natural sources will be upon us within three decades then liquid fuels from biomass could well be at a premium. Moreover, the SRI study on energy crop plantations (see Section 3.1) also presented an apparently powerful case for the conversion of biomass to both SNG and electricity. Comparing biomass with coal and municipal refuse as fuel sources, it has less ash, is low in sulphur content, does not contain the harmful corrosive substances of many refuses, and has a higher energy content than refuse but lower than coal. On the basis of large-scale energy production it is claimed that the efficiency of electrical power generation from the direct combustion of biomass, or via the intermediate preparation of low-energy gases for use in a combined gas turbine – steam cycle (Section 4.5), is 29.4 per cent, which is similar to the efficiency for producing electricity from coal.

With reference to SNG formation, pyrolysis is suggested as being the most suitable conversion technology at present, but that the prospect for the future development of hydrogasification and anaerobic digestion on a large scale could well lead to the eventual selection of either, with anaerobic digestion particularly convenient for aquatic biomass transformations. Based on a dry organic matter to SNG conversion efficiency of 60 per cent, a 2.2 million m^3/day plant with an operating factor of 90 per cent would occupy a land area of some 360 km^2 to yield the required biomass. Water needs would be in the region of 1.5 million m^3/day, approximately 100 times the quantity necessary in SNG from coal enterprises; but additionally ammonia could be produced as a valuable by-product.

More specific ventures worthy of mention include a wood-fired electricity generating station constructed by the Green Mountain Power Company of Vermont, where the annually available renewable wood surplus is thought to be sufficient to fuel a 50 MW plant; the US Union Electric Company and the City of St Louis have jointly operated a power plant combusting 300 t of solid waste per day and generating 12 MW (e) from the recovered energy; the Occidental Research Corporation's flash pyrolysis process dealing with 200 t of refuse/day in San Diego, California; the Union Carbide's Purox system for gasifying 200 t of waste/day at South Charleston, West Virginia; and so the list goes on. At least three states now have wood-fired electricity generating stations, while several more large cities, including Chicago and Nashville, recover energy via urban waste incineration. In 1976 there were upwards of 50 'energy-from-waste' projects either operational or imminently operational within the United States alone having pilot plant, demonstration, or full commercial status. 80 per cent of these involved the treatment of, and materials and energy recovery from, urban sludge and refuse by incineration, pyrolysis, and anaerobic digestion, with one or two examples also of alcoholic fermentation and partial oxidation [157]. The remainder included primarily the anaerobic digestion of livestock manure, with isolated instances of the combustion and pyrolysis of timber and forestry residues and the enzymatic hydrolysis of cellulosic agricultural waste and its subsequent fermentation to ethanol. Twenty-five states were participating in one or more of these schemes, either through an operating company or a particular city authority within the state, and many of the ventures are expected to be relocated and expanded over the next few years. It is to be hoped that this lead will be followed up, though probably on a reduced scale, by other countries in the West, the East, and indeed the Third World, where potentially the gains accruing could and indeed should, be enormous. At the time of writing (mid-1978), the literature emanating from those who genuinely believe that biomass energy production systems will ultimately become vital components of an integrated food, energy, and chemical supply scenario is profuse. This is true as much in Europe [162–166] as in the Americas [167–173]. The challenge is there; the prospects are exciting; the rewards could be great!

7.2 Hydrogen and electricity via biophotolysis – Hope for the future

Many scientists advocate the striving for a situation in which the basic fuel and energy carrier of the future should be hydrogen in the so-called 'hydrogen economy'. Hydrogen can be produced by the 'splitting' of simple water molecules, with the simultaneous formation of oxygen by a variety of methods, including electrolysis, photolysis using a chemical catalyst, and biophotolysis, the topic for discussion here. The advantages of hydrogen over electricity are that it may be stored, and then transported over long distances by pipeline at similar operating costs to electricity transmission. It is a very clean-combusting material, and as a liquid its energy content is around 25 per cent that of petroleum per unit volume but 250 per cent per unit weight. In many industrialized countries there already exist gas distribution networks which can be adapted to the hydrogen economy. As yet only small quantities of gas have been produced utilizing sunlight, and so the efficiencies and scale-up improvements needed remain daunting [174]. This is a futuristic scenario, but a potentially feasible one, and so the basic research being carried on now could yield great dividends, but probably not before well into the next century.

The complex mechanisms of conventional plant photosynthesis were outlined in Section 2.2. It was shown there how the light energy which is absorbed by the plant pigments sets up an initial reaction whereby electrons are removed from water molecules, which themselves dissociate to evolve gaseous oxygen and protons (hydrogen ions). This can be simplified as follows:

$$H_2O \xrightarrow{\text{LIGHT}} 2\,H^+ + [O].$$

Instead of embarking on the reactions of carbon dioxide fixation in the formation of carbohydrates, the released electrons and produced protons can be combined to evolve gaseous hydrogen. This kind of system needs the introduction of hydrogenase enzymes, extracted from blue-green algae like *Spirulina maxima* and *S. platensis* or from non-photosynthetic bacteria such as the experimentally ubiquitous *Escherichia coli* and *Clostridium pasteurianum*, or from certain photosynthetic bacteria, for example, *Chromatium* sp. and *Rhodospirillum rubrum*. The hydrogenase catalyses the linking up of electrons with protons isolated from water to produce minute amounts of hydrogen, just as they do within the cells from which they originated. Thus two main approaches to this kind of research present themselves, one involving the addition of extracted enzyme to isolated choroplasts of plants (spinach being a popular provider), and then illuminating the reaction mixture with a battery of incandescent lamps, and the other by improving the efficiency of hydrogen production by those growing cell cultures which do so naturally.

The first of these has been followed by Professor David Hall and his team at London University, who originally demonstrated the route of electron flow as being from molecular water to the chloroplastic electron carrier protein, ferredoxin, then to the hydrogenase enzyme, before their ultimate incorporation into the liberated hydrogen molecules [175–177]. This group, together with others in a number of countries have succeeded in raising the rates of hydrogen production from modifications of the above generalized system description to about 125 μmoles of H_2/mg chlorophyll/hr from 10 μmoles of H_2/mg chlorophyll/hr after only two years of research. However, the system is sensitive and unstable, hydrogenase is adversely affected by the presence of the evolved oxygen, and a further limiting factor identified is the instability of isolated chloroplast membranes during medium-long periods of continuous illumination.

Improvements have been made, albeit slight ones, in the solving of the above crucial problems and so, in order to make more rapid headway, artificial analogues, specifically synthetic membranes, have been mooted as a means of overcoming the constraints imposed by the purely biological components involved. This concept entails the use of a photoelectric membrane modelled on the basis of the natural photosynthetic membrane, and utilizing chlorophyll as the primary element. Electron transfer would generate a potential difference and develop a current within the external circuit, which could result in the direct production of electricity. This coupling of the initial photosynthetic reactants to electrodes could indeed result in improved energetic efficiencies by, in a sense, by-passing the need to remove and store hydrogen [178]. Whether electricity production would always be a more preferable or profitable product is debatable, but in certain situations this could well be the case and it would obviously be advantageous to provide any system such as this with a degree of flexibility, both in the materials used and in the products formed. Nobel Laureate Professor Melvin Calvin of the University of California anticipates that an artificial membrane system which simulates the basic photosynthesis process, but produces hydrogen at a 75 per cent efficiency of incident light conversion, should be ultimately attainable, and that satisfactory membranes will be developed within 15 years [179].

On the same lines, an intriguing discovery was made a few years ago involving the bacterium *Halobacterium halobium* which, as its name suggests, inhabits salty water and in fact is tolerant to very high salt concentrations. It was observed that this organism was carrying out photosynthesis without the need for chlorophyll, but instead was using a purple pigment, bacteriorhodopsin, which was very similar to the pigment present in the eyes of higher organisms, including man, to aid the conversion of light into electrical impulses – the basis of vision. The mechanism of this type of bacterial photosynthesis is that photons of light are absorbed by the purple pigment and protons are released from within the enclosing membrane. A mini-electrical potential is set up in much the same way as outlined above, which, if the pigment molecules were integrated with a suitable artificial membrane, could be harnessed as an electrical supply for man. The pigment affords a great advantage over the chloroplast pigments of higher plants and algae in that it is surprisingly stable, probably a direct consequence of the extreme ecological conditions which it has to endure when functioning normally within the bacterium itself [180, 181]. Welcome, but in the main unexpected, findings such as this could feasibly constitute a prelude to the major breakthrough which is so obviously needed before any meaningful supply of energy, whether as hydrogen or direct electricity, accrues from this kind of approach.

The fact that actually growing organisms could evolve hydrogen from water was firstly demonstrated by Dr John Benemann and his colleagues at the University of California [182]. The group has since continued with this essentially *in vivo* experimentation on virtually the same nitrogen-fixing blue-green alga, *Anabaena cylindrica*, on which their initial report was based. This filamentous organism has two cell types: vegetative cells and heterocysts. The latter are the sites of nitrogen fixation and are compartmentalized away from the oxygen molecules evolved in photosynthesis. This is because the nitrogenase-mediated reactions are very sensitive to the presence of oxygen. Thus, when hydrogen production is sought the organisms are incubated under an inert gas such as argon, because the nitrogenase enzyme conveniently liberates hydrogen in the absence of molecular nitrogen. Improvements are gradually being manifested in the operation of such a system, the progress being measured by the increasing yields of hydrogen obtained. Originally hydrogen production ensued for three hours from a culture volume of 2 ml at a $H_2 : O_2$ ratio of only 1 : 7; but under conditions of nitrogen starvation the $H_2 : O_2$ ratio has since been improved to 1.7 – 4 : 1 from a 2 litre culture volume incubated over 14 days and more. The latter method has a thermodynamic efficiency of converting incident light energy to hydrogen energy of 0.4 per cent, with possible limiting factors being identified as a decline in the supply of reductant and the physical breakage of the algal filaments [183].

Other nitrogen-fixing algae such as *Nostoc muscorum* have also been used to evolve hydrogen, but apparently not via nitrogenase activity; while the rapidly growing water fern, *Azolla* sp., containing a nitrogen-fixing blue-green algal symbiont, *Anabaena azolla* provides a very favourable system for hydrogen production. By growing the fern in a nitrate medium the *Azolla* becomes independent of the algal nitrogenase, the larger part of the activity of which can then be utilized for hydrogen production generated from water. The nitrogenase enzyme complex is suitably protected from exposure to oxygen in the blue-green alga, which itself can produce both hydrogen and oxygen photosynthetically [184].

Thus, the potential is clearly there for the development of one or more of these schemes, but it remains to be seen how long a period will be required before a technically feasible biophotolysis system is obtained. Succinctly, Joseph Weissman and John Benemann have listed the four principal requirement criteria as being (a) sustained hydrogen and oxygen evolution for many weeks; (b) a 2 : 1 ratio of hydrogen to oxygen; (c) high specific rates of hydrogen production and high photosynthetic efficiencies; (d) no limitations to scale-up of the system [183]. As with many of these situations the problem of scale-up may well prove to be the most critical, but firstly it is important to get the fundamentals as optimum or as close to optimum as possible. At current rates of progress this could certainly be achievable within the not-too-distant future – the question appears to be one of 'when' rather than 'if'?

7.3 Petrol pump plants – A dream come true?

In theory the most logical way of obtaining hydrocarbons from plants is to utilize those species which synthesize them naturally. A great protagonist of this line of thinking is again Melvin Calvin, who, it is claimed, had the idea while waiting two hours in a petrol queue during the advent of the oil crisis in 1973 [185].

The plants in question are species of the genus, *Euphorbia*. Two species in particular, *Euphorbia lathyrus*, a small bush found in northern California, and *Euphorbia tirucalli*, a larger tree and apparently native to Brazil are attracting attention, including that of the Brazilian national petroleum company, Petrobras. *Euphorbia* is related to the giant rubber tree, *Hevea*, but while the latter naturally produces hydrocarbons of molecular weights ranging from 100 000 to a million, the *Euphorbia* latex contains petroleum-like mixtures of molecular weights of only 10 000 to 50 000 [32, 186–188]. Since the bush grows well on arid land, with a minimum requirement for water, the possibility exists for cultivating plantations of *Euphorbia* on land which is marginal for conventional agriculture. Locations in the south-west United States, Australia, Brazil, Iran, and West Africa could well be candidates for these so-called 'petrol plantations'. Replanting would be needed only every 20–30 years. Calvin envisaged an annual crude oil yield of approximately 100 barrels per hectare (13–14 t), but this is before taking into account the formidable energy requirements for fertilization, harvesting, and the reduction of these long-chain compounds to five-or six-carbon atom molecules. The method of extraction would be akin to that for the tapping of rubber trees, except that simultaneous collections would be made from several hundred or thousand trees via a connected pipeline system terminating in a final treatment plant. Calvin himself estimates that the cost of crude hydrocarbons obtained in this manner would be compatible with crude oil at current prices, but on present evidence this seems rather optimistic. It would appear from his claims that *Euphorbia* plantations could surpass the type of biomass plantations discussed in Section 3.1 in most facets, particularly as the latter normally would require subsequent extensive chemical, biological, or heat treatment prior to the production of a fuel available in a utilizable form. Indeed, even if the economics were unfavourable for cultivating the plant as an energy source it could still provide the raw materials necessary to the petrochemical industry as a whole. It should also be possible eventually to manipulate the biosynthetic pathways of certain plants, especially those like *Hevea, Euphorbia*, and the pine tree [189], which produce hydrocarbons naturally, so that the relative proportion of hydrocarbon to carbohydrate formed is significantly increased [190]. This kind of genetic and biochemical engineering is reviewed here in more detail within the following two sections.

A more recent thought, with which Calvin has also been associated, is the concept of a 'hydrogen tree', first publicly proposed by Cesar Marchetti of IIASA, Austria. It seems that certain trees generate hydrogen, which can even today be tapped by the simple expedient of boring a hole.

The yields are trivial, but it has been suggested that, in principle it should be possible to introduce into the tree a blocking reaction which, rather than making the tree produce wood, could allow it to greatly increase its output of hydrogen. This would provide the ideal solar energy-hydrogen route. Marchetti visualizes a forest connected by cheap plastic pipes, leading to a hydrogen main. He estimates that a capital investment of $100 per kW is an attainable figure, comparable to coal-operated plants, and much cheaper than nuclear plants or solar electric systems. However, it must be admitted that for the moment the whole scheme remains speculation. The bioengineering needed for this development does not yet exist, and the fundamental research is barely underway. Indeed, whether such a scheme could ever become a practical proposition must still be a matter for much scepticism.

7.4 Improving plant productivity – I

This section, together with 7.5, is devoted to methods of increasing plant productivity per unit input. Such improvements almost invariably arise from biochemical/genetical techniques acting at the molecular level. The obvious means of achieving higher yields via increased photosynthetic efficiency will be dealt with in the following section, with other methods being discussed here, although the various parameters involved are usually interconnected in some way in any event.

7.4.1 Mixed cropping

The first method under consideration is that of mixed cropping whereby two or more species are cultivated together, in, for example, an energy crop plantation. Higher yields may arise within a given area of land than would have been attained by growing the crops separately but under otherwise apparently identical conditions. This kind of synergistic effect often has no immediate explanation, but the overall better utilization of the solar energy input is an obvious candidate. A certain complementary situation may be envisaged, but the precise nature of such a phenomenon is rarely able to be determined. Greater productivity has been achieved occasionally by cultivating two species side by side out of phase with respect to their growth patterns so as to prolong the period of light interception by quantity of exposed leaf areas, and also by maximizing the leaf area profile so that, for example, upper leaves adapted to higher levels of sunlight and lower leaves to a lesser level could advantageously utilize these differences in combination. Additionally, leaves on tall plant species may preferentially absorb the primary photosynthetic region of the spectrum, while those below are able to utilize light of 'inferior' quality equally well, thus enhancing the combined yield for a given solar energy input. Further research is required in this area, but mixed cropping certainly does offer potential as a way of improving overall yields except, of course, where biomass is required *per se*, irrespective of type or origin, in which case the simultaneous growing of more than one species must produce more biomass than the fastest growing species would in isolation to have any justification [191]. Also to be considered is the possible infection and destruction of a pure monocultural system by external parasites/predators.

7.4.2 Greenhouse cultivation

A second physical, rather than biological or genetic, method of improving energy crop yields could lie in their cultivation within a controlled environment. It is true that greenhouses are rarely used for other than the growth of high quality or exotic crops because of the economics involved, but the widespread construction of inexpensive yet efficient devices remains a possibility, particularly in cold climes and in the arid regions of the south-west USA, Middle East, and Australia. A double film polyethylene structure [192] has been designed to require 30 per cent less energy for heating than a similar sized single-glazed greenhouse, and further energy savings can accrue via the use of selectively reflective curtains and heat-absorbing materials. The roof of such a greenhouse (Figure 7.1) acts as a solar collector during the day, the sun's radiant heat being circulated to a storage tank for night time heating via the medium of 1–2.5 per cent cupric chloride solution, which is transparent to visible light, has a strong absorption in the near infra-red band, and functions as a heat carrier. Recirculation from the tank to the roof releases the heat by night, and in this manner the optimum crop leaf temperature is maintained continuously, thus reducing daytime cooling and ventilation requirements, and diminishing leaf transpiration and hence evaporation [193, 194]. This last factor is a boon to the potential for growing energy crops in areas of limited fresh water supply.

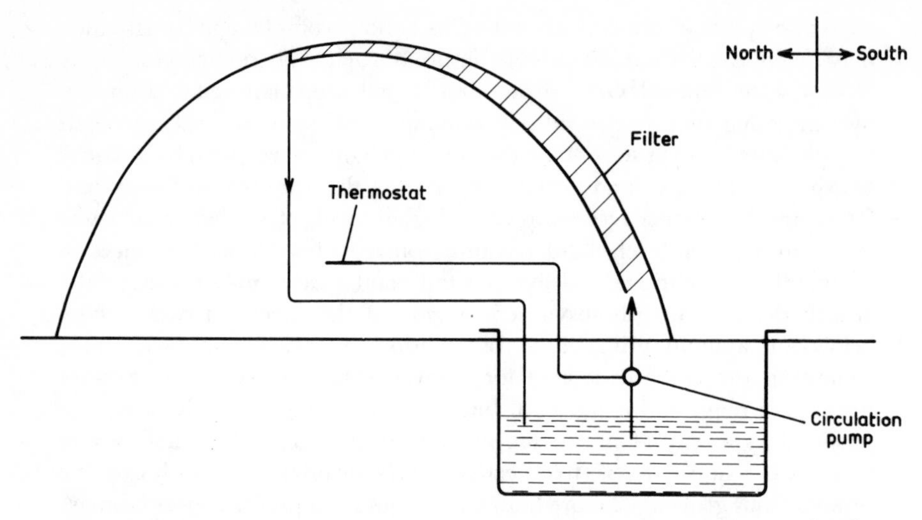

Fig. 7.1 Controlled environment, low energy greenhouse (after Damagnez [193]).

In colder regions the waste heat from power stations and any waste carbon dioxide from industrial processes could both possibly be utilized, the former for raising the internal temperature of the greenhouse and the latter for providing higher levels of potential photosynthate [195]. The use of hydroponic techniques whereby the crops are given access to the minimum quantities of inorganic nutrients and water for maximum yields would probably be applicable, but even with all these multifarious innovations such schemes for growing plants for their energy content alone will almost certainly be uneconomical until the real price of energy is substantially greater than it is at present.

7.4.3 Genetic engineering

Plant genetics offers the possibility of greatly improving plant yields via a number of different ways which will only briefly be outlined here. The production of hybrids using protoplast fusion techniques enables sexually incompatible species, or even genera, to be brought together with resulting greater genetic variations, particularly as problems of cell line regeneration from protoplasts and plant regeneration from cells are gradually being overcome. The use of cell cultures to induce mutations conferring resistance to various diseases including tolerance to extracellular fungal and bacterial toxins, resistance to herbicide application, heavy metals, overproduction of amino acids, and chilling at low temperatures is now on the verge of becoming an extremely valuable tool. Its potential application for selecting out species with improved photosynthetic and reduced photorespiratory levels will be discussed in the following section. Finally, recombinant DNA techniques, whereby desired bacterial genes are integrated into the genome of plant cells, offer the prospect that totally new and hopefully beneficial combinations of genetic material will lead to increased plant productivity. One of the greatest challenges, with possibly the most rewarding prize if successful, lies in the transference of nitrogen fixation genes from bacterial DNA to nonleguminous plants (those unable to form symbiotic associations with nitrogen-fixing organisms). The resulting savings in the use of nitrogenous fertilizers would be a tremendous asset in the striving for high yields with the minimum of energy intensive inputs and make the whole concept of the energy crop plantation a much more attractive one [196].

7.4.4 Nitrogen fixation

At present the only organisms known to be able to fix atmospheric nitrogen are certain members of the procaryotes: bacteria and blue-green algae. The nitrogenase enzyme complex involved catalyses the reduction of nitrogen to ammonia, which is then almost invariably assimilated to amino nitrogen. As well as symbiotic associations between nitrogen-fixing bacteria and leguminous plants there also occurs what is termed associative symbiotic nitrogen fixation in grasses such as maize. In this latter category it is assumed that the limiting factor in nitrogen fixation may be the lack of sufficient carbohydrate supply from the plant to the bacteria. Thus optimum, or at least improved, photosynthesis could reduce this problem and thus extend the range of biological nitrogen-fixing ability. There may well be a trade-off here between production of photosynthate for the plant's own growth and the rate of nitrogen fixation by the bacteria.

As mentioned earlier the introduction of genes coded for nitrogenase production in bacteria into plant cells is now a distinct possibility by virtue of what has become the controversial area of genetic engineering, utilizing recombinant DNA. Already a cluster of nitrogen fixation genes (known as *nif*) has been transferred from the N-fixing bacterium, *Klebsiella pneumoniae* to the non-N-fixing bacterium, *Escherichia coli*. This was accomplished by the sequence of events presented in Figure 7.2 [197] in which the *nif* genes were firstly incorporated into a plasmid (a section of extrachromosomal DNA), before introducing the plasmid into *E. coli* cells. The previously non-N-fixing bacterium was now able to synthesize nitrogenase and fix nitrogen [198]. Nitrogenase enzymes however only function under anaerobic conditions and thus oxygen has to be absent. Free-living N-fixing bacteria like *Klebsiella* sp. do so as anaerobes, while leguminous plants such as the clover and pea possess leghaemoglobin within their root nodules to

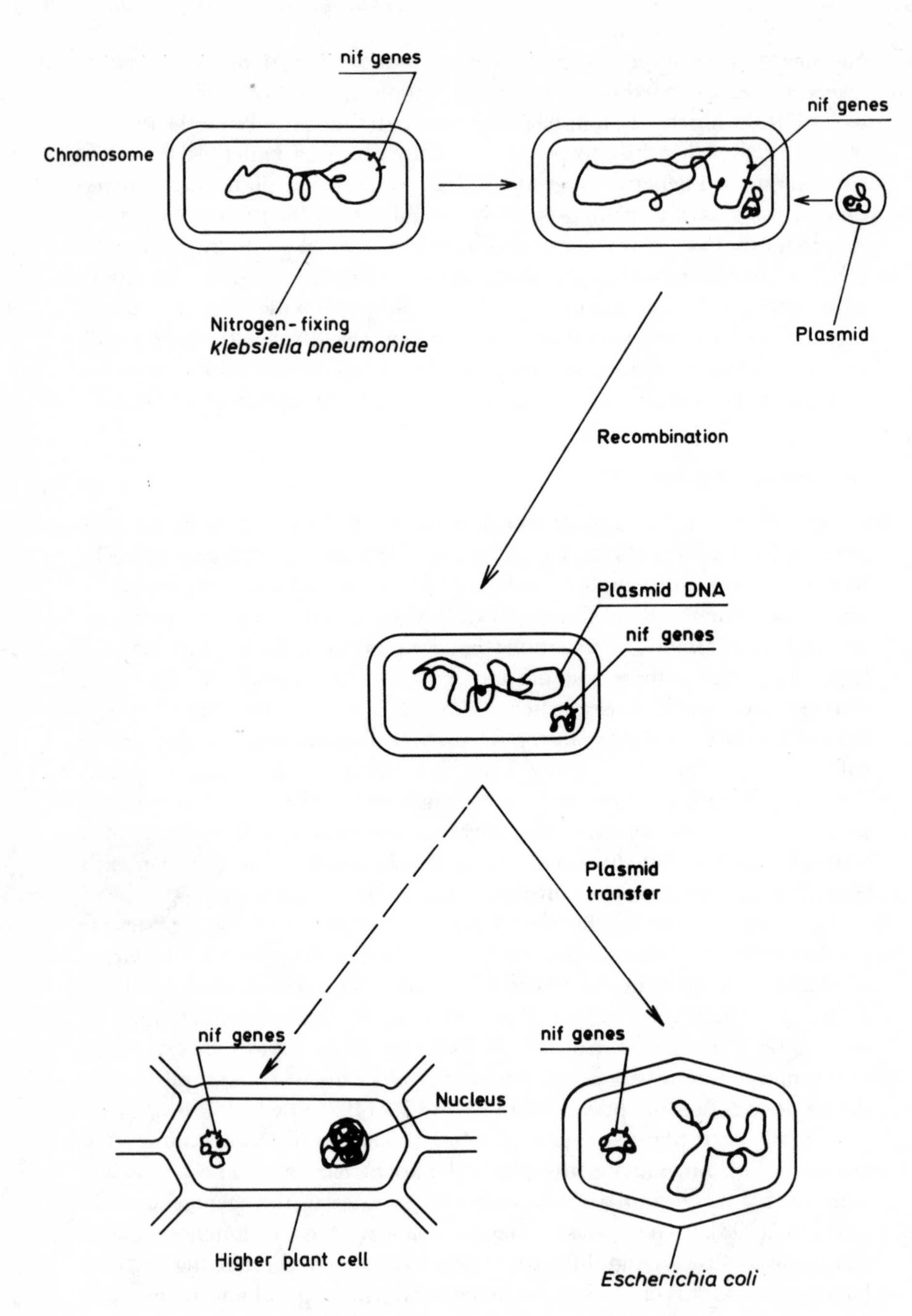

Fig. 7.2 Nitrogen-fixation genes transference.

exclude oxygen from the N-fixing bacterial symbiont, *Rhizobium* sp. This could be the crucial factor in the successful 'manufacturing' of energy crops, or indeed any plants, with the ability to fix their own nitrogen. The protection of the nitrogenase from oxygen is a vital prerequisite of a functioning nitrogen-fixing organism, and so it would seem that other gene clusters controlling a variety of biochemical and physiological parameters would necessarily have to be introduced along with the *nif* genes into a foreign plant host. Thus, although the possibility of producing nitrogen-fixing crops is becoming less of a speculative proposition, many difficulties still need to be met and overcome before it can be considered an absolute reality [199]. These difficulties may, in fact, prove insurmountable.

7.4.5 Metabolic interference

The regulation of plant biosynthetic pathways by genetical/biochemical means to alter the end-products of metabolic sequences could also result in improved yields and energy content in an overall context. One such example concerns the conversion of triose phosphates, either into sucrose for translocation or into pyruvate as a preliminary to the macromolecular synthesis of proteins, fats, etc. It appears that the intracellular concentration of ammonium ions plays a crucial role in the direction taken; either to cellular growth and division or to sucrose export to other parts of the plant. NH_4^+ concentration in turn is possibly controlled by the rate of reduction of NO_3^- in the cytoplasm. Manipulation of the rate of entry of NO_3^- into the cell and its subsequent rate of reduction would, in all probability, unhinge the door controlling the penchant of plant cells towards either sucrose export or protein and fat synthesis at any particular time, and thus open the way for further exploration and potential exploitation [35].

7.5 Improving plant productivity – II

The photosynthetic efficiency of plants has been outlined in Section 1.2, and dealt with in greater depth via the biochemistry involved in Section 2.1. It is obvious that in order to obtain maximum plant productivity, and hence maximum biomass energy output, within a certain area of land (or indeed water), net photosynthesis needs to be optimized; net photosynthesis being equivalent to gross photosynthesis minus respiration.

As explained previously, if CO_2 losses due to respiration could be reduced, then increased net photosynthesis and hence crop yields would result [36], since CO_2 is usually considered to be the critical limiting factor in the photosynthesis of field crops. One method of achieving this could lie in the application of a suitable mutagen to plant tissues and then selecting out cell lines showing enhanced net photosynthesis. Other techniques involve establishing a biochemical basis for genetically altering a specific gene or genes, and are therefore more precise and controllable. Since the oxidation of glycollic acid, an early product of carbon fixation, to glyoxylate is the initial reaction of photorespiration, attempts to block this step should certainly reduce photorespiration considerably [196]. However, the greatest advances in this area thus far have been made concerning the specific regulation of glycollic acid synthesis itself. It has been demonstrated, for instance, that glycidate, an epoxide analogue of glycollate, blocks glycollate synthesis and thus effectively prevents photorespiration in tobacco leaf discs. This results in a 40 per cent increase in net photosynthesis, and additionally the presence of glycidate increases the quantities of other metabolites, namely glutamate and aspartate, within the leaf discs. Furthermore, glutamate and glyoxylate, among several other compounds, have been shown to inhibit the rate of glycollic acid accumulation in the presence of an inhibitor of glycollate oxidase, and this glycollic acid synthesis inhibition in turn is accompanied by a substantial improvement in net photosynthetic CO_2 fixation. These increases constitute around 25 per cent in the presence of glutamate, and about 100 per cent on the blocking of glycollate synthesis and photorespiration with glyoxylate [36]. Thus, genetic mutations affecting the synthesis, utilization and/or accumulation of metabolites such as glutamate and glyoxylate show much promise in decreasing levels of photorespiration, and consequently increasing net photosynthesis and overall biomass productivity. It must be stressed, however, that the genetic engineering required to reduce rates of photorespiration significantly is still some way off, if in fact it is developed at all.

The one possible drawback to such genetic 'tampering' is that no one has yet satisfactorily postulated a positive role for photorespiration, and so its reduction or even elimination could have some hidden deleterious effects on the plant as a whole. Theories have been put forward that it may serve as a source of the amino acids glycine and serine, and/or that it provides possible protection for the plant cell through removal of excess high-energy compounds produced under conditions of high light intensity but low CO_2 levels. This second point can be amplified by thoughts that photorespiration may help to use up excess light energy and so prevent chloroplast damage. By making CO_2 continually available for refixation, photorespiration may contribute to lowering the level of the toxic free-radical superoxide, O_2^-, formed in the electron transference sequence of photosynthesis, so that it can satisfactorily be dealt with by the protective mechanism of each individual chloroplast [200].

Linked with the manifestation of photorespiration is the relationship between C_3 and C_4 plant species, again discussed in Section 2.1. It would appear that a great deal of the difference in net photosynthesis between C_3 and C_4 species can be explained by the slow rate of glycollic acid synthesis, and thus slower photorespiration, in the photosynthetically more efficient C_4 species. Whether a particular plant is of the C_3 or C_4 type is, of course, determined genetically, and it has already been ascertained that the inheritance mechanism of efficient C_4 photosynthesis is complex; and thus far the production of a C_4 hybrid has not been achieved by crossbreeding C_3 with C_4 species. Furthermore, since C_4 plants not only possess an additional biochemical pathway, but also are endowed with a distinctively different anatomy, the production of a formerly C_3 strain possessing C_4 characteristics *in toto* by genetical manipulation would be a difficult proposition indeed!

Better returns could well be obtained by utilizing the C_4 pathway of plants other than by attempted introductions of the pathway into crops that lack it, which may well prove futile. C_4 plants such as maize, sorghum, and especially sugar cane, which are domesticated rather than wild species, could prove to be ideal sources of biomass on marginal lands, particularly in arid or semi-arid environments, if agricultural practice were adjusted to fully exploit their potential. Four-carbon plants are not only highly efficient photosynthetically at high temperatures, but also possess increased potential for the efficient utilization of water during their growth. A second possibility lies in the crossbreeding of domestic species with wild species such as *Tidestromia*. This breeding of C_4 species with other C_4 species could well result in the evolution of hybrids giving higher degrees of productivity than those of the individual parents over a fairly wide range of environmental conditions [33, 201].

In addition to the wastefulness implied in photorespiration is that in dark respiration, which occurs primarily in the mitochondria and which is very similar biochemically to the respiration of animals. Its principal functions are to provide various carbon compounds for utilization in different plant biosynthetic pathways and especially to produce ATP molecules from the reactions of oxidative phosphorylation. Dark respiration occurs as rapidly in the light as in the dark, and values of from 29 per cent to 71 per cent of the gross CO_2 fixed in photosynthesis have been quoted as being used in the dark respiration process. In fact, there are two distinct routes which the respiratory pathway may take in plant mitochondria, the usual one via cytochrome oxidase (which can be blocked by both antimycin A and by cyanide), and an alternative pathway of electron transport which is insensitive to antimycin and cyanide but is specifically inhibited by salicylhydroxamic acid (SHAM), as demonstrated in Figure 7.3. SHAM has no influence

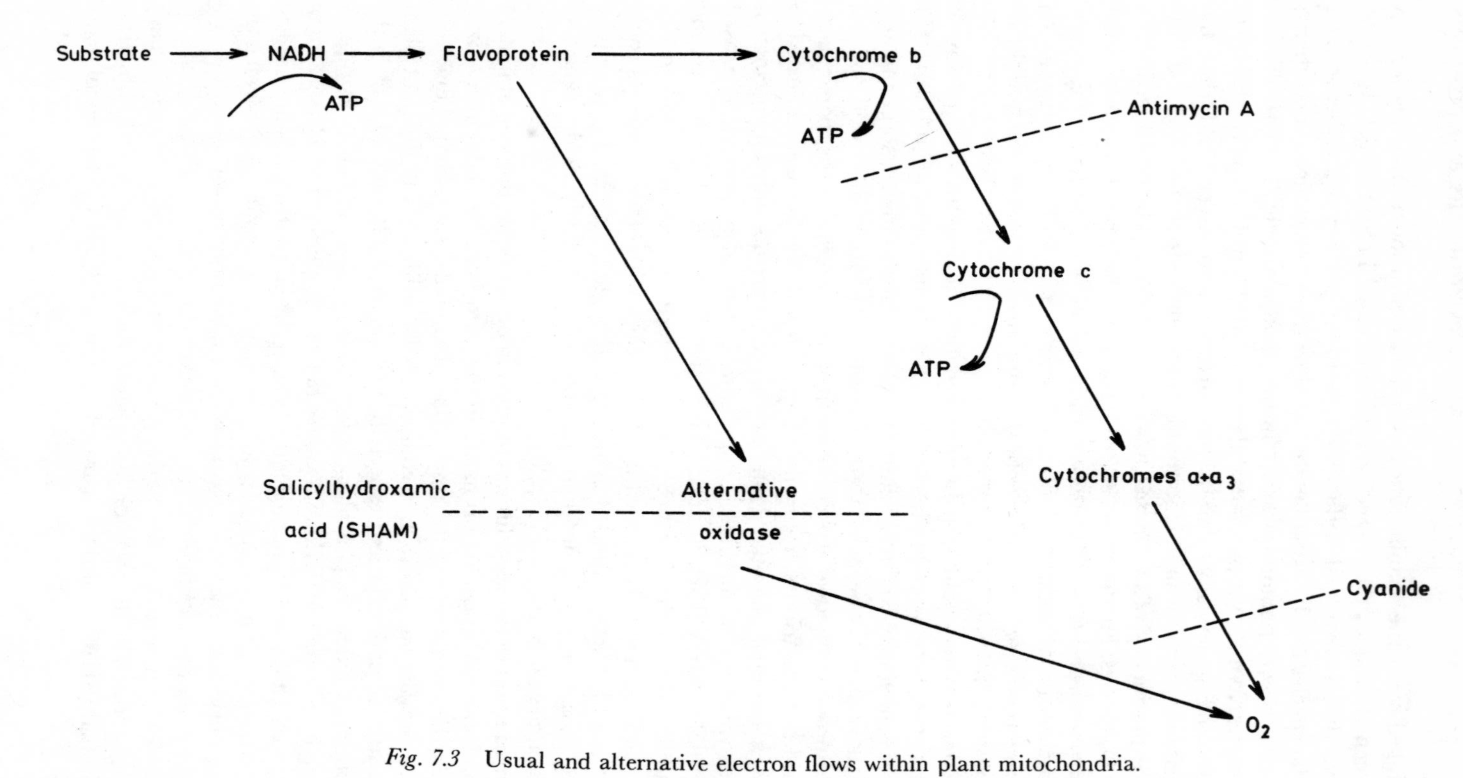

Fig. 7.3 Usual and alternative electron flows within plant mitochondria.

upon the conventional electron transport system. The critical difference between these two routes is that the alternate one produces only one third as many ATP molecules, and thus energy supply, as does the more efficient cytochrome oxidase pathway. There is some doubt as to how much overall photosynthate could actually be retained by the elimination of the energy inefficient pathway, but its contribution to total respiration is normally of the order of from 15–20 per cent. The selection of mutant strains exhibiting tolerance for SHAM, but not for antimycin or cyanide, should not prove too difficult a proposition to accomplish so that, by excluding the wasteful method of dark respiration, an increase in plant productivity would be achieved [36]. Nevertheless, as with attempts to reduce photorespiration, success can as yet by no means be taken for granted.

In summary, Section 7.4 together with 7.5, attempt to pinpoint the main areas where plant productivity can be enhanced through man's increasing, but by no means omniscient, awareness of the intricacies of molecular biology, genetics, biochemistry, and physiology. Much remains to be elucidated, but ideas that were not so long ago mere speculations of men with fertile minds could soon be translated into improved crop yields on land with sometimes less than fertile soil.

7.6 Ligno-cellulose enzyme hydrolysis – Energy-saving biochemistry

In Sections 4.3 and 5.5 the necessity of ball-milling, an energy-intensive treatment, for rendering cellulose and ligno-cellulosic materials amenable to subsequent enzymatic attack was emphasized. Since cellulose is the most abundant carbohydrate available for bioconversion, then it is important that a technique be developed to treat the ligno-cellulose complex both more economically and by using less energy.

The most advanced research into this problem has been carried out over the last 10 years by Dr Karl-Erik Eriksson at the Swedish Forest Products Research Laboratory in Stockholm [202]. Working mainly with wood as substrate at least 11 enzymes involved in the degradation of cellulose have been identified, along with two connected with lignin breakdown. These enzymes are extracellular and are produced by the white rot fungus, *Sporotrichum pulverulentum*, a thermo-tolerant organism with an optimum growth temperature of 39 °C. Thus, although the cellulose enzymes are by now well documented, knowledge of those which are present in the lignin degradation route is still somewhat scarce. Additionally, whereas the cellulosic breakdown enzyme system(s) have largely been demonstrated *in vitro*, that of lignin breakdown has thus far been confined to *in vivo* studies.

It has been established that crystalline cellulose is split to form monosaccharide and oligosaccharide sugars, as well as acids, by the combined action of endo- and exo- 1,4-β-glucanases, the former of which attack β-1,4-glucosidic bonds (see Section 4.3) indiscriminately throughout the cellulose polymer. In fact, five different endo-1,4-β-glucanase enzymes have been isolated from *S. pulverulentum*. The function of the exo-glucanase is to split off glucose and cellobiose units from the non-reducing end of a chain, with a simultaneous inversion of the product configurations from the β- to the α-form. Moreover, a recently discovered cellobiose oxidase enzyme is thought to play a role in these reactions, either in a regulatory capacity or possibly to oxidise reducing end-groups produced on cleavage of a β-1,4-glucosidic bond.

The subsequent oxidation of the disaccharide cellobiose proceeds to give eventually, cellobionic acid, either via the action of cellobiose oxidase or via a lignin-dependent route, whereby the reduction of quinones and/or phenoxy radicals also occurs, mediated by a cellobiose : quinone oxido-reductase. These latter moieties are formed by the oxidation of lignin phenols, which themselves repress cellulose enzyme production, by laccase or peroxidase enzymes. A β-1,4- glucosidase converts both cellobiose and cellobionic acid to glucose, and gluconic acid plus glucose lactone, respectively. The last of these compounds inhibits β-glucosidase, and monosaccharides present in high concentrations also result in catabolite repression of the endo-glucanases. Further inhibitions of various enzymes are assumed to occur in a rather complicated manner, but the net result is that fermentable sugars are produced from the normally recalcitrant ligno-cellulosic complex. As yet the rates of reaction are slow, however, and ways must be found for increasing these before the system could be incorporated into a large-scale commercial operation.

7.7 Conservation through integration – A systems approach

It is perhaps important to stress that although, for convenience and ease of reference, individual topics have often been dealt with in isolation throughout this book, the crux of any fuels-from-biomass system is that it should be one of integration. There should, as far as possible, be a completely harmonious relationship between the raw material inputs, the process(es), the primary end-product(s), any by-product(s) formed, and the surrounding environment. The elimination of all wastes would be an obvious priority here and, as it happens, many bioconversion systems also perform a waste-treatment function at the same time, as has been noted beforehand. Professor Oswald's algal-bacterial system for sewage treatment allied to solar energy fixation and subsequent methanogenesis (Section 4.2, Figures 4.3 and 4.4) provides an excellent example of such an integrated system.

Pursuing a similar theme, an example is taken here of an investigation into the possibilities for converting solar energy into biological energy, methane energy, and ultimately electrical energy in sequence. The

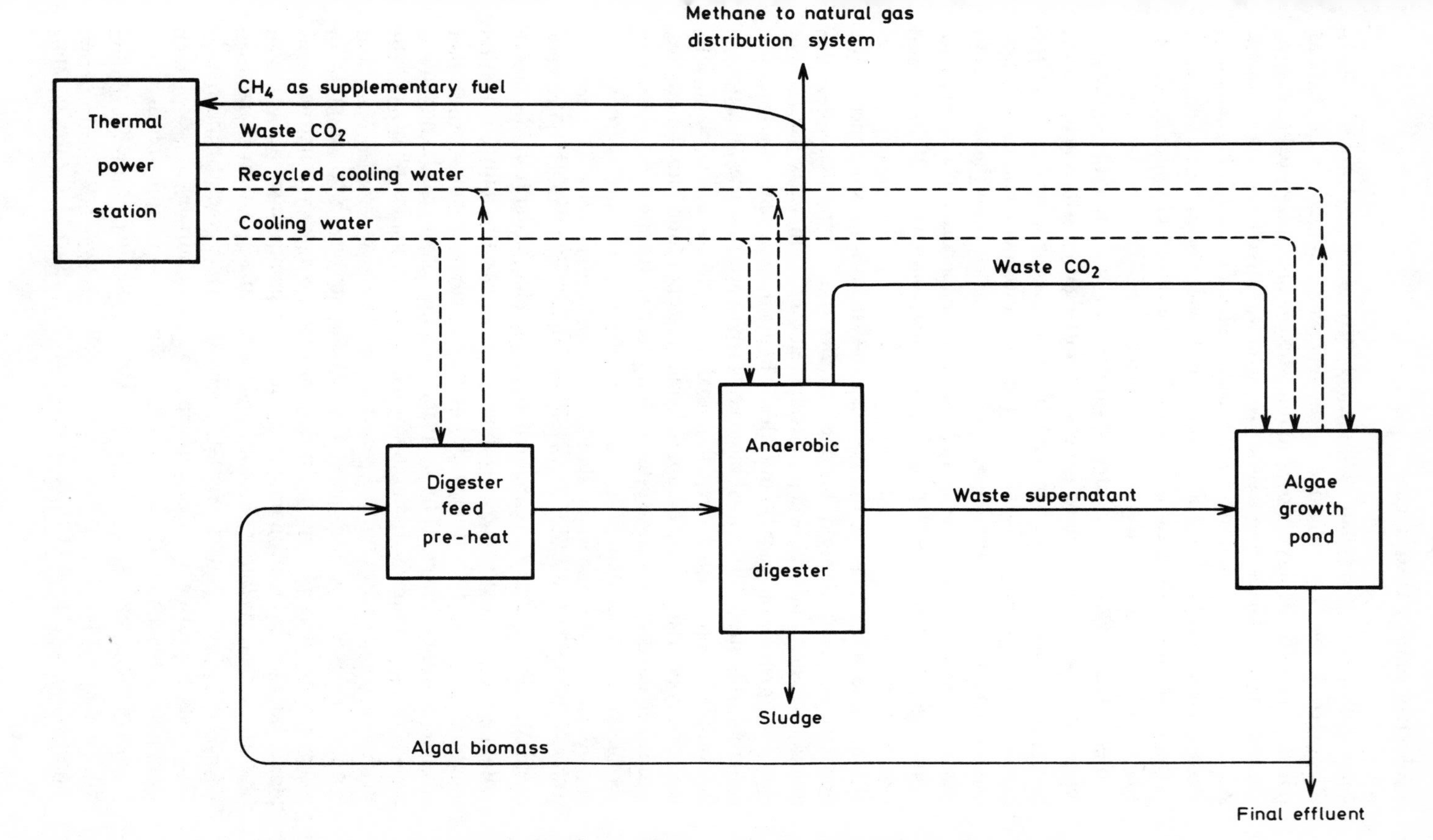

Fig. 7.4 Integrated algae-to-methane system (after Keenan [203]).

particular study by Keenan of the University of Pennsylvania [203], is considered to be representative of an increasing number of postulated schemes currently being proposed in this whole area of integrated biological systems analysis. Great emphasis is placed on waste recovery and recycling, as may be observed from the diagram presented as Figure 7.4. For instance, waste carbon dioxide and cooling water are recycled from the power plant, to the algal cultivation pond in the case of the former and to the pond and anaerobic digester in that of the latter. The carbon dioxide aids photosynthesis, while the water is utilized to maintain the optimum temperature rises for methane generation and algal growth. Furthermore, waste supernatant containing valuable nutrients is recycled from the digester to the pond along with waste carbon dioxide, again to aid algal productivity, while the cooling water is also recycled from the digester to the power station condenser to fulfil its cooling function once more. The primary source of nutrients for algal growth, too, could well be derived from wastewaters.

The mode of operation of the various sub-systems are, as outlined in the previous sections, devoted to photosynthesis, algal productivity, and methane generation. However, some relevant comments concerning electricity generation may be in order. The methane emanating from the anaerobic digestion of the algal biomass is envisaged to act as an auxiliary fuel supply for a principally fossil fuel-fired steam-electric power plant. This methane may alternatively be upgraded for incorporation into the existing natural gas distribution system; but, concentrating on electricity production, the expedient utilization of methane evolved largely from solar energy and waste materials should improve the present 33 per cent efficiency for fossil-fuel power plants by reducing the fossil fuel requirement itself. Additionally, the recycling of the waste cooling water from the power station provides something of a bonus by at least partially alleviating potential hazards of thermal pollution of the environment and possible adverse ecological effects. The need for provision of several large cooling towers is also diminished, thus improving the economics of the overall scheme as compared to normal power station practice.

Small-scale experiments have been performed by Keenan and his colleagues in which they have determined the energy efficiencies of photosynthesis by the filamentous, nitrogen-fixing, alkaline tolerant, blue-green alga, *Anabaena flos-aquae* and of the subsequent methanogenic stage. The values obtained were allied to data from practical large-scale operations appearing in the literature, to arrive initially at potentially attainable conversion efficiency targets under optimum biochemical engineering management conditions. Thus, assuming a photosynthetic efficiency at 4.5 per cent (which is admittedly optimistic but nonetheless conceivable), insolation at 16.75 MJ/m^2 day, and 16.75 MJ/kg dry organic matter energy content, a projected biomass crop yield of 60 t dry organic matter/ha year is envisaged to be achievable. However, for the purposes of presenting a plausible hypothesis, various parameters of photosynthetic efficiency, biomass harvesting energy demand and efficiency, digestion efficiency, and the supplementary energy demand for functions such as mixing, pumping and processing were studied prior to the derivation of a so-called 'typical' or 'nominal' scenario.

They calculated that 4088 ha of *Anabaena flos-aquae* cultivated with a reasonable photosynthetic efficiency of 2.3 per cent and at a solar energy flux of 16.75 MJ/m^2 day produce 300 t of dry organic biomass per 24 hours.

This material is daily transferred to the digesters, although the assumption made for the harvesting energy requirement at 10 per cent of the subsequent energy content of the methane may be considered on the low side. The difficulties of harvesting algal cells at a low energy expenditure have been noted in previous sections, although allowances may be made in the case of these floating filamentous types. At an anticipated conversion efficiency of biomass energy to methane energy content of 60 per cent, 7.45 TJ of methane fuel would be evolved each day, of which the equivalent of 745 GJ, 84 GJ, 50 GJ, and 5 GJ are deemed to have been expended on biomass harvesting, pumping, mixing, and post-digestion processing respectively. This represents a very favourable energy out : fossil fuel energy input ratio of 8.4 : 1. This is conceivable because of the waste energy and nutrient recovery and recycling operations within the overall system mentioned above. In terms of overall solar energy conversion to methane, a very satisfactory 1.1 per cent efficiency is attained. This is equal to a net energy output of 160 GJ/ha year, equivalent to 100 000 barrels of oil per year from the 4088 ha area of cultivation.

This is just an example of what might be achieved through innovative thinking combined with an awareness of the established facts. It is a reasonably 'down-to-earth' appraisal of what can be done within certain quite well-defined limits, although it is true that some of the assumptions of energy requirements and efficiencies may be slightly optimistic. The principles are sound and form a basis for a larger scale investigation to ascertain how close the hypothetical calculations are to the absolute truth. This is the only method by which such schemes can be tried and tested, and only then can they be put to use, not only for the benefit of mankind, but also for the benefit of the biosphere as a whole.

7.8 Biomass potential for national energy autonomy

It is an interesting and revealing exercise to examine the potential for energy provision from biomass systems from a national perspective. This potential is related to two main factors: national energy density (expressed as energy consumed/ha of land), and biomass productivity (expressed as net energy output, GJ/ha year).

Below, we illustrate what percentage of land area would have to be devoted to biomass production in each of three nations for a good, but nevertheless realizable, net energy of 65 GJ/ha yr (10 barrels of oil/ha yr).

Country	*Area (km^2)*	*1973 Energy use (GJ)*	*Energy density (GJ/ha)*	*Percent land area to provide demand*
Rhodesia	375 000	10×10^7	2.65	4 percent
Austria	81 800	32×10^7	38.5	60 percent
Belgium	29 400	110×10^7	380	580 percent

The less intensively developed, less intensively populated countries clearly have the greatest potential for meeting their energy demands via biomass sources. However, the situation is obviously modified by factors such as climate, terrain, and land prices.

7.9 The biological path to self-reliance

As seen in Section 7.8, bioenergy production systems have a role to play in the energy system of the industrialized world, but most of this supply will probably accrue from the importation of liquid fuels derived from biomass sources originating in the tropics. Not only might these 'biofuels' serve as energy providers by way of a substitute for petroleum [204], but also as natural chemical feedstocks, vital for the expansion of national economies.

However, it is in the Third World countries themselves where the potential impact of biomass systems could be greatest. Indeed, the logical outcome of a book of this nature is to consider whether a thus far ill-developed community can utilize biosystems to 'boot-strap' itself into a higher level of development. This huge question has, in fact, been addressed by the International Federation of Institutes for Advanced Study (IFIAS) under its project leader, Professor Carl-Gören Heden. Expressed at its simplest, the idea is that by the application of modern bioenergy and agricultural techniques a village community could intensify its output without recourse to industrial inputs, and so break what has become the tragedy of the poverty trap.

For example, one could visualize a situation in which the community was given an initial contribution, from a more prosperous country, of a certain amount of technology and, over a period of years, a continuing supply of scientific advice. One step might be the introduction of a Gobar gas plant, the recycle of the soil conditioner from that gas plant on to the fields, the growth in these fields of leguminous plants, part of which being ploughed back into the land to raise the nitrogen level. The ensuing gradual build-up of the animal stock and thus of more waste, and the eventual introduction of algal ponds could soon follow. Little of this, of course, refers to bioenergies; yet as the system develops, the need for energy and energy sources rises and part of the existing terrain of the community could be switched to bioenergy producing systems. Such a study must be looked at in a systematic way and IFIAS has a proposal for a dynamic systems analysis of such a community to be tested on village communities spread throughout Asia and South America. The objective is to demonstrate that this route of development is neither so slow nor so primitive that enterprising Third World countries would reject it as being a second-best approach to their national development. Such is the belief of the leaders of this project in the potential benefits of microbiological and bioenergetic processes that they consider these real time, real life experiments will have the effect of demonstrating to Third World leaders that there is a viable, indeed preferable, alternative route to national development beyond simply copying the energy-intensive systems of the West, based as they are, upon fissile and fossil non-renewable fuel sources.

References

[1] Lovins, A. B. (1977), *Soft Energy Paths*, Penguin Books, Harmondsworth.

[2] Hall, D. O. (1974), Presented at *International Solar Energy Society Inaugural Meeting—UK Section*, London, January 1974.

[3] Bassham, J. A. (1976), *Clean Fuels from Biomass, Sewage, Urban Refuse, Agricultural Wastes* p. 205, Institute of Gas Technology, Chicago.

[4] Dyer, A. D. (1976), *History Today* p. 598, September 1976.

[5] Stevens, H. M. (1963), *The Advancement of Science*, **20**, 1.

[6] Eckholm, E. P. (1975), *Natural History* p. 7, October 1975.

[7] Reed, T. B. and Lerner, R. M. (1973), *Science*, **182**, 1299.

[8] Grayson, A. J. (1974), *Wood Resources and Demands: A Statistical Review*, HMSO, London.

[9] Openshaw, K. (1974), *New Scientist*, **61**, 271.

[10] Miller, D. L. (1975), *Biotechnol. Bioeng. Symp. No. 5*. p. 345, John Wiley and Sons, New York, London, Sydney and Toronto.

[11] Rose, D. (1976), *Process Biochemistry*, **11**(2), 10.

[12] White, D. (1976), *Financial Times News Features*, 24 March 1976, London.

[13] Egloff, S. (1938), *Ind. Eng. Chem.*, **30**, 1091.

[14] Saeman, J. F. and Andreasen, A. A. (1954) *Industrial Fermentations Vol. 1* (eds. L. A. Underkofler and R. J. Hickey) p. 136, Chemical Publishing Co., New York.

[15] Marshall, J. E., Petrick, G. and Chan, H. (1975). *Canadian Forestry Service Information Report E-X-25*, Ottawa.

[16] Goldstein, I. S. (1975), *Science*, **189**, 847.

[17] Texeira, C., Andreasen, A. A. and Kolachov, P. (1950), *Ind. Eng. Chem.*, **42**, 1781.
[18] Jackson, E. A. (1976), *Process Biochemistry*, **11**(5), 29.
[19] Goldemberg, J. (1978), *Science*, **200**, 158.
[20] de Carvalho Jr., A. V., Milfont, Jr., W. N., Yang, V. and Trindade, S.C. (1977), Presented at *Int. Symp of Alcohol Fuel Technology – Methanol and Ethanol*, Wolfsburg, FRG.
[21] Yang, V., Milfont Jr., W. N., Scigliano, A., Massa, C. O., Sresnewsky, S. and Trindade, S. C. (1977), Presented at *12th Intersociety Energy Conversion Engineering Conf., Washington D.C.*
[22] Buswell, A. M. (1954), *Industrial Fermentations Vol. 2* (eds. L. A. Underkofler and R. J. Hickey) p. 518, Chemical Publishing Co., New York.
[23] Summers, R. and Bousfield, S. (1976), *Process Biochemistry*, **11**(5), 3.
[24] UK–ISES (1976), *Solar Energy, – a UK Assessment*, Int. Solar Energy Soc. – UK Section, London.
[25] Duffie, J. A. and Beckman, W. A. (1974), *Solar Thermal Energy Processes*, John Wiley and Sons, New York, London, Sydney and Toronto.
[26] Szeicz, G. (1974), *J. Appl. Ecology*, **11**, 617.
[27] Dept. of Energy (1976), *Solar Energy: Its Potential Contribution within the United Kingdom*, HMSO, London.
[28] Brinkworth, B. J. (1972), *Solar Energy for Man*, The Compton Press, Salisbury.
[29] McMullan, J. T., Morgan, R. and Murray, R. B. (1976), *Energy Resources and Supply*, John Wiley and Sons, New York, London, Sydney and Toronto.
[30] Hall, D. O. and Rao, K. K. (1977), *Photosynthesis*, 2nd edn., Edward Arnold, London.
[31] Hall, D. O. and Whatley, F. R., (1967), *Enzyme Cytology* (ed. D. B. Roodyn) Chap. 4. Academic Press, London.
[32] Calvin, M. (1976) *Photochem.Photobiol.*, **23**, 425.
[33] Bjorkman, O. and Berry, J. (1973), *Sci. Am.*, **229**, (4), 80.
[34] Marx, J. L. (1973), *Science*, **179**, 365.
[35] Bassham, J. A. (1977), *Science*, **197**, 630.
[36] Zelitch, I. (1975), *Science*, **188**, 626.
[37] Zelitch, I. (1977), *European Seminar on Biological Solar Energy Conversion Systems* L-4, Grenoble-Autrans, May 1977.
[38] de Kouchkovsky, Y. (1977), *European Seminar on Biological Solar Energy Conversion Systems* L-2, Grenoble-Autrans, May 1977.
[39] Chartier, P. (1977), *European Seminar on Biological Solar Energy Conversion Systems* L-3, Grenoble-Autrans, May 1977.
[40] Box, E. (1975), *Primary Productivity of the Biosphere* (eds. H. Lieth and R. H. Whittaker), p. 265, Springer-Verlag, New York.
[41] Cooper, J. P. (1975), *Photosynthesis and Productivity in Different Environments* (ed. J. P. Cooper) p. 593, Cambridge University Press.

[42] Cooper, J. P. (1976), *Food Production and Consumption: The Efficiency of Human Food Chains and Nutrient Cycles* (eds. A. N. Duckham, J. G. W. Jones and E. H. Roberts) p. 107, North Holland Publishing Co.
[43] Hall, D. O. (1976), Presented at *CENTO Solar Energy Conf.*, Tehran, 3–5 October 1976.
[44] Loomis, R. S. and Gerakis, P. A. (1975), *Photosynthesis and Productivity in Different Environments* (ed. J. P. Cooper) p. 145, Cambridge University Press.
[45] Alich Jr., J. A. and Inman, R. E. (1976), *Clean Fuels from Biomass, Sewage, Urban Refuse, Agricultural Wastes*, p. 287, Institute Of Gas Technology, Chicago.
[46] Inman, R. E. (1975), *Biotechnol. Bioeng. Symp. No. 5* p. 67, John Wiley and Sons, New York, London, Sydney and Toronto.
[47] Alich Jr., J. A. and Inman, R. E. (1976), *Energy*, **1**, 53.
[48] Inman, R. E. (1977), *Silvicultural Biomass Farms Vol. 1—Summary*, The Mitre Corporation/Metrek Division, USA.
[49] Howlett, K. and Gamache, A. (1977), *Silvicultural Biomass Farms Vol. 2 – The Biomass Potential of Short-Rotation Farms*, The Mitre Corporation/Metrek Division, USA.
[50] Salo, D. J., Inman, R. E., McGurk, B. J. and Verhoeff, J. (1977), *Silvicultural Biomass Farms Vol. 3 – Land Suitability and Availability*, The Mitre Corporation/Metrek Division, USA.
[51] Inman, R. E., Salo, D. J. and McGurk, B. J. (1977), *Silvicultural Biomass Farms Vol. 4 – Site Specific Production Studies and Cost Analyses*, The Mitre Corporation/Metrek Division, USA.
[52] Bliss, C. and Blake, D. O. (1977), *Silvicultural Biomass Farms Vol. 5 – Conversion Processes and Costs*, The Mitre Corporation/Metrek Division, USA.
[53] Howlett, K. and Gamache, A. (1977), *Silvicultural Biomass Farms Vol. 6 – Forest and Mill Residues as Potential Sources of Biomass*, The Mitre Corporation/Metrek Division, USA.
[54] Inman, R. E. (1977), *European Seminar on Biological Solar Energy Conversion Systems* L-7, Grenoble-Autrans, May 1977.
[55] Golueke, C. G., Oswald, W. J. and Gotaas, H. B. (1957), *Appl. Microbiol.*, **5**, 47.
[56] Oswald, W. J. (1960), *J. Sanit. Eng. Div. Proc. Am. Soc. Civ. Eng.*, **86**, No. SA4, 7.
[57] Golueke, C. G. and Oswald, W. J. (1963), *Solar Energy*, **7** (3), 86.
[58] Oswald, W. J. and Golueke, C. G. (1968), *Single-Cell Protein* (eds. R. I. Mateles and S. R. Tannenbaum) p. 271, M.I.T. Press, Cambridge, Mass. and London.
[59] Oswald, W. J. (1973), *Solar Energy*, **15**, 107.
[60] Uziel, M., Oswald, W. J. and Golueke, C. G. (1975), *Solar Energy Fixation and Conversion with Algal-Bacterial Systems* (Final Project Report, National Science Foundation Grant No. Gl-39216), University of California, Berkeley.

[61] Kosaric, N., Nguyen, H. T. and Bergougnou, M. A. (1974), *Biotechnol. Bioeng.*, **16**, 881.
[62] Benemann, J. R., Weissman, J. C., Koopman, B. L. and Oswald, W. J. (1977), *Nature*, **268**, 19.
[63] Oswald, W. J. and Benemann, J. R., *Proc. of U.S. – Japan Conf. on Photosynthetic Energy Conversion*, Academic Press, New York (in press).
[64] Benemann, J. R., Koopman, B. L. and Oswald, W. J. (1977), Presented at *Fuels from Biomass Symp.*, Univ. of Illinois, Champaign, 18–19 April 1977.
[65] Lapointe, B. E., Williams, L. D., Goldman, J. C. and Ryther, J. H. (1976), *Aquaculture*, **8**, 9.
[66] Goldman, J. C. and Ryther, J. H. (1976), Presented at *Biological Solar Energy Conversion Conf.*, Miami, 15–18 November 1976.
[67] Goldman, J. C. and Ryther, J. H. (1976), *Biological Control of Water Pollution* (eds. J. Tourbier and R. W. Pierson) p. 197, Univ. of Pennsylvania Press.
[68] Goldman, J. C. and Ryther, J. H. (1977), *Biological Solar Energy Conversion* (eds. A. Mitsui, S. Miyachi, A. San Pietro and S. Tamura) p. 367, Academic Press, New York.
[69] Goldman, J. C. (1978), *Fuels from Solar Energy: Photosynthetic Systems – State of the Art and Potential for Energy Production* (Final Report, Dept of Energy Contract No. EG-77-S-02-4151), Woods Hole Oceanographic Institution, Mass.
[70] Leese, T. M. (1976), *Clean Fuels from Biomass, Sewage, Urban Refuse, Agricultural Wastes*, p. 253, Institute of Gas Technology, Chicago.
[71] Wilcox, H. (1976), *Proc. of Conf. on Capturing the Sun through Bioconversion* p. 255, Washington D.C., 10–12 March 1976.
[72] Wolverton, W. and McDonald, R. C. (1976), *New Scientist*, **71**, 318.
[73] Lecuyer, R. P. (1976), *Clean Fuels from Biomass, Sewage, Urban Refuse, Agricultural Wastes* p. 267, Institute of Gas Technology, Chicago.
[74] Bellamy, W. D. (1974), *Biotechnol. Bioeng.*, **16**, 869.
[75] Stutzenberger, F. J., Kaufman, A. J. and Lossin, R. D. (1970), *Can. J. Microbiol.*, **16**, 553.
[76] Porteous, A. (1977), *Recycling Resources Refuse*, Longman Group, London.
[77] Atchison, J. E. (1976), *Science*, **191**, 768.
[78] Heslop-Harrison, J. (1975), *Biologist*, **22**, 60.
[79] Raymond, W. F. (1977), *Biologist*, **24**, 80.
[80] United Nations (1974), *Statistical Yearbook 1973*, United Nations, New York.
[81] Miller, D. L. (1973), *Industrial Uses of Cereals* (chairman Y. Pomeranz) p. 252, Am. Ass. of Cereal Chemists Inc., St. Paul, Minnesota.
[82] Gaden, E. L. (1974), *Single Cell Protein* (ed. P. Davis) p. 47, Academic Press, London, New York and San Francisco.

[83] Smith, D. L. O., Rutherford, I. and Radley, R. W. (1975), *The Agricultural Engineer*, **30**(3), 70.
[84] Saddler, H. D. W. (1975), *Organic Wastes and Energy Crops as Potential Sources of Fuel in Australia*, Univ. of Sydney.
[85] Horton, R. and Hawkes, D. (1976), *Energy World*, **28**, 3.
[86] Hobson, P. N., Robertson, A. M. and Mills, P. J. (1975), *Agricultural Research Council Research Review*, **1**(3), 82.
[87] Hawkes, D. Horton, R. and Stafford, D. A. (1976), *Process Biochemistry*, **11**(3), 32.
[88] Horton, R., Hawkes, D. F. and Bassett, M. B. (1975), *The Heating and Ventilating Engineer*, **49**, 14.
[89] Prasad, C. R., Prasad, K. K. and Reddy, A. K. N. (1974), *Econ. Polit. Weekly*, **IX**, 32–34, Special Number, Bombay, August 1974.
[90] Parikh, J. K. and Parikh, K. S. (1977), *Energy*, **2**, 441.
[91] Anon (1975), *New Scientist*, **67**, 655.
[92] Smith, F. A. (1977). Personal communication.
[93] Holdom, R. S. (1976), Personal communication.
[94] Oswald, W. J. (1976), *Clean Fuels from Biomas, Sewage, Urban Refuse, Agricultural Wastes*, p. 311, Institute of Gas Technology, Chicago.
[95] Oswald, W. J. (1976), Presented at *Symp. on Energy Bioconversion*, Campinas, Sao Paulo, Brazil, 8 July 1976.
[96] Hobson, P. N. (1973), *Process Biochemistry*, **8** (1), 19.
[97] Finney, C. D. and Evans, R. S. (1975), *Science*, **190**, 1088.
[98] Evans, W. C. (1977), *Nature*, **270**, 17.
[99] Wise, D. L., Sadek, S. E., Kispert, R. G., Anderson, L. C. and Walker, D. H. (1975), *Biotechnol. Bioeng. Symp. No. 5.* p. 285, John Wiley and Sons, New York, London, Sydney and Toronto.
[100] Loll, U. (1976), *Microbial Energy Conversion* (eds. H. G. Schlegel and J. Barnea) p. 361, Erich Goltze, KG., Gottingen, FRG.
[101] Oshima, M. (1965), *Wood Chemistry, Process Engineering Aspects* p. 26, Noyes Development Corporation, (cited by ref. 52 pB-2).
[102] Harrison, J. S. and Graham, J. C. J. (1970), *The Yeasts Vol. 3.* (eds. A. H. Rose and J. S. Harrison) p. 283, Academic Press, London and New York.
[103] Conn, E. E. and Stumpf, P. K. (1972), *Outlines of Biochemistry*, 3rd edn., John Wiley and Sons, New York, London, Sydney and Toronto.
[104] Stanier, R. Y., Doudoroff, M. and Adelberg, E. A. (1971), *General Microbiology*, 3rd edn., The Macmillan Press, London.
[105] Burnett, J. H., (1968), *Fundamentals of Mycology*, Edward Arnold, London.
[106] Prescott, S. C. and Dunn, C. G. (1959), *Industrial Microbiology*, 3rd edn., p. 858, McGraw-Hill, New York, Toronto and London.
[107] Yamada, N. and Tomoda, K. (1966), US Patent, 3, 293, 142.
[108] Underkofler, L. A. (1954), *Industrial Fermentations Vol. 2* (eds. L. A. Underkofler and R. J. Hickey) p. 97, Chemical Publishing Co., New York.

[109] Fleming, I. D. (1968), *Starch and its Derivatives* (by J. A. Radley) p. 498, Chapman and Hall, London.
[110] Worgan, J. T. (1973), *The Biological Efficiency of Protein Production* (ed. J. G. W. Jones), p. 339, Cambridge University Press.
[111] Spano, L. A., Medeiros, J. and Mandels, M. (1975), *Enzymatic Hydrolysis of Cellulosic Wastes to Glucose* (Report of US Army Natick Development Center, Mass.).
[112] Nystrom, J. M. and Allen, A. L. (1975), Presented at *Symp. on Enzymatic Conversion of Cellulosic Materials*, Newton, Mass., 8 September 1975.
[113] Nystrom, J. M. (1975), *Biotechnol. Bioeng. Symp. No. 5* p. 221, John Wiley and Sons, New York, London, Sydney and Toronto.
[114] Wilke, C. R. and Mitra, G. (1975), *Biotechnol. Bioeng. Symp. No. 5* p. 253, John Wiley and Sons, New York, London, Sydney and Toronto.
[115] Katz, M. and Reese, E. T. (1968), *Appl. Microbiol.*, **16**, 419.
[116] Brandt, D., Hontz, L. and Mandels, M. (1973), *A.I.Ch.E. Symp. Series*, **69**, 133.
[117] Brandt, D, (1975), *Biotechnol. Bioeng. Symp. No. 5* p. 275, John Wiley and Sons, New York, London, Sydney and Toronto.
[118] Ghose, T. K. (1969), *Biotechnol. Bioeng.*, **11**, 239.
[119] Ghose, T. K. and Kostick J. A. (1970), *Biotechnol. Bioeng.*, **12**, 921.
[120] Mandels, M., Hontz, L. and Nystrom, J. (1974), *Biotechnol. Bioeng.*, **16**, 1471.
[121] Union Carbide Corporation (1977), (Cited by ref. 52, p. 78).
[122] Preston, G. T. (1976), *Clean Fuels from Biomass, Sewage, Urban Refuse, Agricultural Wastes* p. 89, Institute of Gas Technology, Chicago.
[123] Knight, J. A. Bowen, M. D. and Purdy, K. R. (1976), Presented at *Conf. of the Energy and Wood Products Industry and Forest Products Research Society*, Atlanta, Georgia, 15–17 November 1976.
[124] Dept. of Energy (1977), *District Heating Combined with Electricity Generation in the United Kingdom*, HMSO, London.
[125] Earl, D. E., (1975) *Forest Energy and Economic Development*, Clarendon Press, Oxford.
[126] Slesser, M., (1978), *Energy in the Economy*, The Macmillan Press, London and New York.
[127] IFIAS, (1974), *Workshop Report N. 6*, International Federation of Institutes for Advanced Study, Solna, Sweden.
[128] McCann, D. J. and Saddler, H. D. W., (1976), *Search*, **1–2**, 17.
[129] McCann, D. J. and Saddler, H. D. W., (1976), *J. Austral. Inst. Agric. Sci.* p. 41, March 1976.
[130] Saddler, H. D. W., McCann, D. J. and Pitman, M. G. (1976), *Australian Forestry*, **39**, 5.
[131] Lewis, C. W. (1977), *Energy*, **2**, 241.
[132] Trindade, S. C. (1978), Personal communication.
[133] Blaxter, K. L. (1975), *Biologist*, **22**, 14.

[134] Leach, G. (1975), *Food Policy*, **1**, 62.
[135] Leach, G. (1976), *Energy and Food Production*, IPC Press, Guildford.
[136] Chandran, T. C., (1976), Personal communication.
[137] Holdom, R. S., Winstrom–Olsen, B. and Cocker, R. (1977), *European Seminar on Biological Solar Energy Conversion Systems* A–27, Grenoble–Autrans, May 1977.
[138] Pimentel, D., Nafus, D., Vergara, W., Papaj, D., Jaconetta, L., Wulfe, M., Olsvig, L., Frech., K., Loye, M., and Mendoza, E. (1978), *Bio Science* **28**, 376.
[139] Anon (1976), *Ecos* **9**, 21 (Citing G. Gartside).
[140] Hammond, A. L. (1977), *Science*, **195**, 564.
[141] Lewis, C. W., *Biochemical and Photosynthetic Aspects of Energy Production* (ed. A. San Pietro), Academic Press (in Press).
[142] Lewis, C. W. (1976), *J. Appl. Chem. Biotechnol.* **26**, 568.
[143] Morse, R. N. and Siemon, J. R. (1976), Presented at *Annual Engineering Conference*, Inst. of Engineers, Townsville, Australia.
[144] Szokolay, S. V. (1976) *Solar Energy and Building*, The Architectural Press, London.
[145] Dept. of Energy (1977), *Compare your Home Heating Costs*, HMSO, London.
[146] Slesser, M. and Hounam, I. (1976), *Nature*, **262**, 244.
[147] Szego, G. (1976), *Capturing the Sun through Bioconversion* p. 217, Washington Center for Metropolitan Studies, Washington DC.
[148] Fraser, M. D., Henry, J. F., and Vail, C. W. (1976), *Clean Fuels from Biomass, Sewage, Urban Refuse, Agricultural Wastes* p. 371, Institute of Gas Technology, Chicago.
[149] Roller, W. L., Keener, H. M., Kline, R. D., Mederski, H. J. and Curry, R. B. (1975), *Grown Organic Matter as a Fuel Raw Material Resource*, National Aeronautics and Space Administration Report No. NASA–CR–2608, Washington DC.
[150] Langerhorst, J., Prast, G. and Thalhammer, T. (1977), *Solar Energy – Report on a Study of the Difficulties Involved in Applying Solar Energy in Developing Countries* (Prepared for the Minister for Development Cooperation, The Netherlands, 11 February 1977).
[151] Alich, Jr., J. A., Schooley, F. A., Ernest, R. K., Miller, K. A., Lonks, B. B., Veblen, T C., Witwer, J. G. and Hamilton, R. H. (1977), *European Seminar on Biological Solar Energy Conversion Systems* L–8, Grenoble–Autrans, May 1977.
[152] Rexen, F. (1977). Personal communication.
[153] Boer, K. W. (1978) *Solar Energy*, **20**, 25.
[154] Ryle, M. (1977), *Nature*, **267**, 111.
[155] Glendenning, I. and Count, B. M. (1977), *Proc. Symp. Renewable Sources of Energy*, Roy. Soc. Arts. p. 50.
[156] Hammond, A. L. and Metz, W. D. (1977), *Science*, **197**, 241.
[157] Anon (1976), *Chem. Eng. News*, **54** (8), 24.
[158] Hall, D. O. and Slesser, M. (1976), *New Scientist*, **71**, 136.

[159] Johansson, T. B. and Steen, P. (1978), *Ambio*, **7**, 70.
[160] Hall, D. O. and Lalor, E. (1977), Presented at *European Seminar on Biological Solar Energy Conversion Systems*, Grenoble–Autrans, May 1977.
[161] Anon (1977), *New Scientist*, **75**, 666.
[162] Hasdenteufel, J. B. (1978), *2nd International Solar Forum Vol II* p. 17, Deutsche Gesellschaft fur Sonnenenergie, Hamburg, July 1978.
[163] Hall, D. O. (1978), *Bio-Energy; Energy from Living Systems* p. 26, Gottlieb Duttweiler–Institut, Zurich.
[164] Svikonsky, E. (1978), *Bio-Energy; Energy from Living Systems*, p. 127, Gottlieb Duttweiler–Institut, Zurich.
[165] Heden, C-G., *Bioenergetic Aspects of Microbial Processes*, Biochem, Soc. Transactions (in press).
[166] Williams, L. A., Foo, E. L., Foo, A. S., Kuhn, I. and Heden C-G. (1978) Presented at *Symp. on Biotechnology in Energy Production and Conservation*, Catlinburg, Tennessee, 10–12 May 1978.
[167] Lipinsky, E. S. (1978), *Science*, **199**, 644.
[168] Anon (1978), *Science*, **200**, 636.
[169] Wishart, R. S. (1978), *Science*, **199**, 614.
[170] Calvin, M. (1978), *Bio-Energy: Energy from Living Systems* p. 89, Gottlieb Duttweiler–Institut, Zurich.
[171] Gaddy, J. L. (1978), *Bio-Energy; Energy from Living Systems* p. 184, Gottlieb Duttweiler–Institut, Zurich.
[172] Szego, G. C., Fraser, M. D., and Henry, J-F. (1978), *2nd International Solar Forum Vol II* p. 1, Deutsche Gesellschaft für Sonnenenergie, Hamburg, July 1978.
[173] Klass, D. L. (1976), *Clean Fuels from Biomass, Sewage, Urban Refuse, Agricultural Wastes*, p. 21, Institute of Gas Technology, Chicago.
[174] Bylinsky, G. (1976), *Fortune*, September 1976, 152.
[175] Hall, D. O., Rao, K. K., Reeves, S. G., Adams, M. W. W., and Dennis, G. (1977), *European Seminar on Biological Solar Energy Conversion Systems* L–24, Grenoble–Autrans, May 1977.
[176] Rao, K. K., Rosa, L. and Hall, D. O. (1976), *Biochem. Biophys. Res. Comm.*, **68**, 21.
[177] Reeves, S. C., Rao, K. K., Rosa, L. and Hall, D. O. (1976), *Microbial Energy Conversion* (eds. H. G. Schlegel and J. A. Barnea) p. 235, Erick Goltze, KG, Gottingen, FRG.
[178] San Pietro, A. (1977), *European Seminar on Biological Solar Energy Conversion Systems* L–23, Grenoble–Autrans, May 1977.
[179] Hammond, A. L. (1977), *Science*, **197**, 745.
[180] Anon (1976), *New Scientist*, **69**, 547.
[181] Oesterhelt, D. (1977), *European Seminar on Biological Solar Energy Conversion Systems* L–26, Grenoble–Autrans, May 1977.
[182] Benemann, J. R., and Weare, N. M. (1974), *Science*, **184**, 174.
[183] Weissman, J. C. and Benemann, J. R. (1977), *Appl. Environ. Microbial.*, **33**, 123.

[184] Newton, J. W. (1976), *Science,* **191**, 559.
[185] *The Guardian* (1977), 1 February 1977, London and Manchester.
[186] *The Guardian* (1977), 9 December, 1977, London and Manchester.
[187] Anon (1976), *Science,* **194**, 46.
[188] Nielsen, P. E., Nishimura H., Otvos, J. W. and Calvin, M. (1977), *Science*, **198**, 942.
[189] Anon (1970), *Technology Review*, **79**, 17.
[190] Roberts, D. P. (1976), *Capturing the Sun through Bioconversion* p. 673, Washington Center for Metropolitan Studies, Washington DC.
[191] Willey, R. W. and Roberts, E. H. (1976), *Solar Energy in Agriculture* p. 44, UK–ISES Conf. C9, London.
[192] Roberts, W. J. and Mears, D. R. (1976), *Solar Energy in Agriculture* p. 69, UK–ISES Conf. C9, London.
[193] Damagnez, J. (1976), *Solar Energy in Agriculture* p. 82, UK–ISES Conf. C9, London.
[194] Damagnez, J., Chiapale, J. P. and Denis, P. (1977), *European Seminar on Biological Solar Energy Conversion Systems* L–11, Grenoble–Autrans, May 1977.
[195] Siddall, R. G. (1978), *Solar Energy for Industry* p. 59, UK–ISES Conf. C9, London.
[196] Day, P. R (1977), *Science*, **197**, 1334.
[197] Brill, W. J. (1977), *Sci. American*, March 1977, 68.
[198] Dixon, R. A. (1977), *European Seminar on Biological Solar Energy Conversion Systems L–21, Grenoble–Autrans, May 1977.*
[199] Dixon, R. A. (1978), *New Scientist*, **78**, 684.
[200] Barber, J. and Halliwell, B. (1977), *Nature*, **270**, 104.
[201] Wittwer, S. H. (1974), *Bio Science*, **24**, 216.
[202] Erikkson, K–E. (1977), *European Seminar on Biological Solar Energy Conversion Systems* L–22, Grenoble–Autrans, May 1977.
[203] Keenan, J. D. (1977), *Energy,* **2**, 365.
[204] Spurgeon, D. (1978), *Nature*, **273**, 702.

Appendix 1

Conversion factors and orders of magnitude

ENERGY

1 MJ = 0.27778 kWh = 238.85 kcal = 947.92 BTU = 0.37251 hph.

1 tce (ton coal equivalent) = 26.9 GJ

1 tpe (tonne petroleum equivalent) = 7.5 barrels (North Sea Crude) = 44.8 GJ

1000 ft^3 natural gas = 28.3 m^3 natural gas = 1.1 GJ

AREA

1 hectare = 2.47 acres = 10^4 m^2

VOLUME

1 m^3 = 1000 litres = 35.315 ft^3 = 219.97 UK gallons = 264.17 US gallons

MASS

1 kg = 2.205 lbs = 0.001 tonne (t)

1 tonne = 0.984 tons = 1.102 short tons = 1000 kg = 2205 lbs

POWER

1 W = 1 Js^{-1} = 0.001 kW = 0.001341 hp

PRESSURE

1 Pa (Pascal) = 1 newton m^{-2} (Nm^{-2}) = 0.10197 $kgfm^{-2}$ = 0.00014504 $lbftin^{-2}$

ORDERS OF MAGNITUDE

kilo = 10^3; mega = 10^6; giga = 10^9; tera = 10^{12}; peta = 10^{15}; exa = 10^{18}; micro(μ) = 10^{-6}; nano(n) = 10^{-9}

Appendix 2

Gross energy requirements (GERs) of selected inputs

(Values given are time and technology specific and relate to latest technology in 1976).

Input	*Unit*	*MJ*
Electricity (from fuel-fired power stations)	kWh	14
Fuel oil	li(kg)	44.5 (46.3)
Natural gas (methane)	m^3(kg)	38.8 (25.5)
Process water	m^3 (t)	2.0
Process steam	kg	3.4
Sulphuric acid	kg	0
Urea	kg	36.0
Ammonia (gas)	kg	50.6
Ammonia (liquid)	kg	48.3
$(NH_4)_2SO_4$	kg	14.5
H_3PO_4	kg	5.7
P	kg	14.0
K	kg	9.7
Organic fungicides, herbicides, pesticides	kg	approx 120
Stainless steel (vessel)	kg	91
Stainless steel (pipe)	kg	68
Structural steel	kg	50
Concrete	kg	9.5
$MgSO_4$	kg	9.0
Antifoam	kg	34.0

Appendix 3

Computation of present value

(No increase of energy prices through time)

$i = 0.0$

YEARS	0.01	0.02	0.03	0.04	0.05	0.06	0.07	0.08	0.09	0.10	0.11	0.12	0.13	0.14	0.15
	r value														
1	0.990	0.980	0.971	0.962	0.952	0.943	0.935	0.926	0.917	0.909	0.901	0.893	0.885	0.877	0.870
2	1.970	1.942	1.913	1.886	1.859	1.833	1.808	1.783	1.759	1.736	1.713	1.690	1.668	1.647	1.626
3	2.941	2.884	2.829	2.775	2.723	2.673	2.624	2.577	2.531	2.487	2.444	2.402	2.361	2.322	2.283
4	3.902	3.808	3.717	3.630	3.546	3.465	3.387	3.312	3.240	3.170	3.102	3.037	2.974	2.914	2.855
5	4.853	4.713	4.580	4.452	4.329	4.212	4.100	3.993	3.890	3.791	3.696	3.605	3.517	3.433	3.352
6	5.795	5.601	5.417	5.242	5.076	4.917	4.767	4.623	4.486	4.355	4.231	4.111	3.998	3.889	3.784
7	6.728	6.472	6.230	6.002	5.786	5.582	5.389	5.206	5.033	4.868	4.712	4.564	4.423	4.288	4.160
8	7.651	7.325	7.020	6.733	6.463	6.210	5.971	5.747	5.535	5.335	5.146	4.968	4.799	4.639	4.487
9	8.565	8.162	7.786	7.435	7.108	6.802	6.515	6.247	5.995	5.759	5.537	5.328	5.132	4.946	4.772
10	9.471	8.982	8.530	8.111	7.722	7.360	7.024	6.710	6.418	6.145	5.889	5.650	5.426	5.216	5.019
11	10.367	9.787	9.253	8.760	8.306	7.887	7.499	7.139	6.805	6.495	6.207	5.938	5.687	5.453	5.234
12	11.254	10.575	9.954	9.385	8.863	8.384	7.943	7.536	7.161	6.814	6.492	6.194	5.918	5.660	5.421
13	12.133	11.348	10.635	9.986	9.393	8.853	8.358	7.904	7.487	7.103	6.750	6.424	6.122	5.842	5.583
14	13.003	12.106	11.296	10.563	9.899	9.295	8.745	8.244	7.786	7.367	6.982	6.628	6.302	6.002	5.724
15	13.864	12.849	11.938	11.118	10.380	9.712	9.108	8.559	8.061	7.606	7.191	6.811	6.462	6.142	5.847
16	14.717	13.577	12.561	11.652	10.838	10.106	9.447	8.851	8.313	7.824	7.379	6.974	6.604	6.265	5.854
17	15.561	14.292	13.166	12.166	11.274	10.477	9.763	9.122	8.544	8.022	7.549	7.120	6.729	6.373	6.047
18	16.397	14.992	13.753	12.659	11.689	10.828	10.059	9.372	8.756	8.201	7.702	7.250	6.840	6.467	6.128
19	17.225	15.678	14.324	13.134	12.085	11.158	10.336	9.604	8.950	8.365	7.839	7.366	6.938	6.550	6.198
20	18.045	16.351	14.877	13.590	12.462	11.470	10.594	9.818	9.129	8.514	7.963	7.469	7.025	6.623	6.259
21	18.856	17.011	15.415	14.029	12.821	11.764	10.836	10.017	9.292	8.649	8.075	7.562	7.102	6.687	6.312
22	19.659	17.658	15.937	14.451	13.163	12.042	11.061	10.201	9.442	8.772	8.176	7.645	7.170	6.743	6.359
23	20.455	18.292	16.444	14.857	13.488	12.303	11.272	10.371	9.580	8.883	8.266	7.718	7.230	6.792	6.399
24	21.242	18.914	16.935	15.247	13.799	12.550	11.469	10.529	9.707	8.985	8.348	7.784	7.283	6.835	6.434
25	22.022	19.523	17.413	15.622	14.094	12.783	11.654	10.675	9.823	9.077	8.422	7.843	7.330	6.873	6.464
26	22.794	20.121	17.877	15.983	14.375	13.003	11.826	10.810	9.929	9.161	8.488	7.896	7.372	6.906	6.491
27	23.558	20.707	18.327	16.330	14.643	13.210	11.987	10.935	10.027	9.237	8.548	7.943	7.409	6.935	6.514
28	24.315	21.281	18.764	16.663	14.898	13.406	12.137	11.051	10.116	9.307	8.602	7.984	7.441	6.961	6.534
29	25.064	21.844	19.188	16.984	15.141	13.591	12.278	11.158	10.198	9.370	8.650	8.022	7.470	6.983	6.551
30	25.806	22.396	19.600	17.292	15.372	13.765	12.409	11.258	10.274	9.427	8.694	8.055	7.496	7.008	6.566
31	26.541	22.937	20.000	17.588	15.593	13.929	12.532	11.350	10.343	9.479	8.733	8.085	7.518	7.020	6.579
32	27.268	23.468	20.389	17.874	15.803	14.084	12.647	11.435	10.406	9.526	8.769	8.112	7.538	7.035	6.591
33	27.988	23.988	20.766	18.148	16.002	14.230	12.754	11.514	10.464	9.569	8.801	8.135	7.556	7.048	6.600
34	28.701	24.498	21.132	18.411	16.193	14.368	12.854	11.587	10.518	9.609	8.829	8.157	7.572	7.060	6.609
35	29.407	24.998	21.487	18.665	16.374	14.498	12.948	11.655	10.567	9.644	8.855	8.176	7.586	7.070	6.617
36	30.106	25.488	21.832	18.908	16.547	14.621	13.035	11.717	10.612	9.677	8.879	8.192	7.598	7.079	6.623
37	30.798	25.969	22.167	19.143	16.711	14.737	13.117	11.775	10.653	9.706	8.900	8.208	7.609	7.087	6.629
38	31.483	26.440	22.492	19.368	16.868	14.846	13.193	11.829	10.691	9.733	8.919	8.221	7.618	7.094	6.634
39	32.161	26.902	22.808	19.584	17.017	14.949	13.265	11.879	10.726	9.757	8.936	8.233	7.627	7.100	6.638
40	32.833	27.355	23.115	19.793	17.159	15.046	13.332	11.925	10.757	9.779	8.951	8.244	7.634	7.105	6.642
41	33.498	27.799	23.412	19.993	17.294	15.138	13.394	11.967	10.787	9.799	8.965	8.253	7.641	7.110	6.645
42	34.156	28.234	23.701	20.186	17.423	15.225	13.452	12.007	10.813	9.817	8.977	8.262	7.647	7.114	6.648
43	34.808	28.661	23.982	20.371	17.546	15.306	13.507	12.043	10.838	9.834	8.989	8.270	7.652	7.117	6.650
44	35.454	29.080	24.254	20.549	17.663	15.383	13.558	12.077	10.861	9.849	8.999	8.276	7.657	7.120	6.652
45	36.093	29.490	24.519	20.720	17.774	15.456	13.606	12.108	10.881	9.863	9.008	8.283	7.661	7.123	6.654
46	36.725	29.892	24.775	20.885	17.880	15.524	13.650	12.137	10.900	9.875	9.016	8.288	7.664	7.126	6.656
47	37.352	30.286	25.025	21.043	17.981	15.589	13.692	12.164	10.918	9.887	9.024	8.293	7.668	7.128	6.657
48	37.972	30.673	25.267	21.195	18.077	15.650	13.730	12.189	10.934	9.897	9.030	8.297	7.671	7.130	6.659
49	38.586	31.052	25.502	21.341	18.169	15.708	13.767	12.212	10.948	9.906	9.036	8.301	7.673	7.131	6.660
50	39.194	31.423	25.730	21.482	18.256	15.762	13.801	12.233	10.962	9.915	9.042	8.305	7.675	7.133	6.661

Appendix 3 contd. (3%/year rate of increase of energy prices)

$i = 0.03$

	r value														
YEARS	0.0	0.01	0.02	0.04	0.05	0.06	0.07	0.08	0.09	0.10	0.11	0.12	0.13	0.14	0.15
1	1.030	1.020	1.010	0.990	0.981	0.972	0.963	0.954	0.945	0.936	0.928	0.920	0.911	0.904	0.896
2	2.091	2.060	2.029	1.971	1.943	1.916	1.889	1.863	1.838	1.813	1.789	1.765	1.742	1.720	1.698
3	3.184	3.120	3.059	2.943	2.887	2.833	2.781	2.731	2.682	2.634	2.588	2.543	2.500	2.457	2.416
4	4.309	4.202	4.099	3.905	3.813	3.725	3.640	3.558	3.479	3.403	3.329	3.258	3.190	3.124	3.060
5	5.468	5.305	5.149	4.858	4.721	4.591	4.466	4.347	4.232	4.123	4.017	3.916	3.819	3.726	3.636
6	6.662	6.430	6.209	5.801	5.612	5.433	5.262	5.099	4.944	4.797	4.656	4.521	4.393	4.270	4.152
7	7.892	7.577	7.280	6.736	6.486	6.251	6.028	5.817	5.617	5.428	5.248	5.078	4.915	4.761	4.615
8	9.159	8.747	8.361	7.662	7.344	7.046	6.765	6.501	6.253	6.019	5.798	5.589	5.392	5.205	5.029
9	10.464	9.940	9.452	8.578	8.185	7.818	7.475	7.154	6.854	6.572	6.308	6.060	5.826	5.607	5.400
10	11.807	11.156	10.555	9.486	9.010	8.568	8.158	7.777	7.421	7.090	6.781	6.492	6.222	5.969	5.732
11	13.192	12.397	11.668	10.386	9.819	9.297	8.816	8.370	7.958	7.575	7.220	6.890	6.583	6.297	6.030
12	14.617	13.662	12.792	11.276	10.613	10.006	9.449	8.937	8.465	8.030	7.628	7.256	6.912	6.593	6.296
13	16.086	14.953	13.927	12.158	11.392	10.695	10.058	9.477	8.944	8.455	8.006	7.593	7.212	6.860	6.535
14	17.598	16.269	15.073	13.032	12.156	11.364	10.645	9.991	9.396	8.853	8.357	7.902	7.485	7.102	6.748
15	19.156	17.610	16.231	13.897	12.905	12.014	11.210	10.483	9.824	9.226	8.683	8.187	7.734	7.320	6.940
16	20.761	18.979	17.400	14.754	13.640	12.645	11.753	10.951	10.228	9.576	8.985	8.449	7.961	7.517	7.111
17	22.414	20.375	18.580	15.602	14.361	13.259	12.276	11.398	10.610	9.903	9.265	8.689	8.168	7.695	7.265
18	24.116	21.798	19.772	16.442	15.069	13.856	12.780	11.824	10.971	10.209	9.525	8.911	8.357	7.856	7.403
19	25.869	23.249	20.975	17.275	15.763	14.435	13.265	12.230	11.312	10.496	9.767	9.114	8.529	8.002	7.526
20	27.675	24.729	22.191	18.099	16.443	14.998	13.732	12.618	11.634	10.764	9.991	9.302	8.686	8.133	7.636
21	29.536	26.239	23.418	18.915	17.111	15.546	14.181	12.987	11.939	11.015	10.199	9.474	8.828	8.252	7.735
22	31.452	27.778	24.657	19.724	17.766	16.077	14.613	13.340	12.227	11.251	10.392	9.632	8.959	8.359	7.823
23	33.425	29.348	25.909	20.525	18.409	16.594	15.030	13.676	12.499	11.471	10.571	9.778	9.077	8.456	7.903
24	35.458	30.949	27.173	21.318	19.039	17.096	15.430	13.996	12.756	11.678	10.737	9.912	9.186	8.544	7.974
25	37.551	32.582	28.449	22.103	19.657	17.584	15.816	14.302	12.998	11.871	10.891	10.035	9.284	8.623	8.037
26	39.708	34.247	29.737	22.881	20.264	18.058	16.188	14.594	13.228	12.052	11.034	10.148	9.374	8.694	8.094
27	41.929	35.945	31.039	23.652	20.859	18.518	16.545	14.872	13.445	12.221	11.166	10.252	9.456	8.759	8.145
28	44.217	37.676	32.353	24.414	21.442	18.966	16.889	15.137	13.650	12.380	11.290	10.348	9.531	8.817	8.191
29	46.573	39.442	33.680	25.170	22.015	19.401	17.220	15.390	13.843	12.528	11.404	10.436	9.599	8.870	8.232
30	49.001	41.243	35.020	25.919	22.577	19.824	17.539	15.631	14.026	12.667	11.510	10.517	9.661	8.918	8.269
31	51.500	43.079	36.373	26.660	23.127	20.234	17.846	15.861	14.199	12.798	11.608	10.592	9.717	8.961	8.302
32	54.075	44.952	37.739	27.394	23.668	20.633	18.142	16.080	14.362	12.920	11.700	10.660	9.769	8.999	8.331
33	56.728	46.862	39.119	28.121	24.198	21.021	18.426	16.290	14.517	13.034	11.784	10.723	9.816	9.035	8.357
34	59.459	48.810	40.512	28.841	24.718	21.398	18.700	16.489	14.663	13.141	11.863	10.781	9.859	9.066	8.381
35	62.273	50.796	41.919	29.554	25.228	21.764	18.963	16.680	14.800	13.241	11.936	10.835	9.898	9.095	8.402
36	65.171	52.822	43.339	30.260	25.729	22.120	19.217	16.861	14.931	13.335	12.004	10.884	9.933	9.121	8.421
37	68.156	54.888	44.774	30.960	26.219	22.465	19.461	17.034	15.054	13.423	12.066	10.929	9.966	9.144	8.438
38	71.231	56.994	46.223	31.652	26.701	22.801	19.696	17.199	15.170	13.505	12.125	10.970	9.995	9.166	8.453
39	74.398	59.143	47.685	32.338	27.173	23.128	19.923	17.357	15.280	13.582	12.179	11.008	10.022	9.185	8.467
40	77.659	61.334	49.163	33.018	27.637	23.445	20.141	17.507	15.384	13.654	12.229	11.043	10.047	9.202	8.479
41	81.019	63.568	50.654	33.691	28.091	23.753	20.350	17.650	15.482	13.721	12.275	11.075	10.069	9.218	8.490
42	84.480	65.846	52.161	34.357	28.537	24.052	20.552	17.787	15.575	13.784	12.319	11.105	10.090	9.232	8.499
43	88.044	68.170	53.682	35.017	28.974	24.343	20.746	17.917	15.662	13.844	12.359	11.132	10.108	9.244	8.508
44	91.715	70.540	55.218	35.671	29.403	24.626	20.934	18.041	15.745	13.899	12.396	11.157	10.125	9.256	8.516
45	95.497	72.956	56.769	36.318	29.824	24.901	21.114	18.160	15.823	13.951	12.430	11.181	10.141	9.266	8.523
46	99.391	75.421	58.335	36.960	30.237	25.168	21.287	18.273	15.897	13.999	12.463	11.202	10.155	9.276	8.529
47	103.403	77.934	59.917	37.595	30.642	25.427	21.454	18.380	15.967	14.045	12.492	11.221	10.168	9.284	8.535
48	107.535	80.497	61.514	38.223	31.040	25.679	21.614	18.483	16.033	14.088	12.520	11.239	10.179	9.292	8.540
49	111.791	83.111	63.127	38.846	31.429	25.924	21.769	18.581	16.096	14.127	12.545	11.256	10.190	9.299	8.545
50	116.174	85.777	64.755	39.463	31.812	26.162	21.918	18.675	16.155	14.165	12.569	11.271	10.200	9.305	8.549

$i = 0.10$

YEARS	0.0	0.01	0.02	0.03	0.04	0.05	0.06	0.07	0.08	0.09	0.11	0.12	0.13	0.14	0.15
						r value									
1	1.100	1.089	1.078	1.068	1.058	1.048	1.038	1.028	1.018	1.009	0.991	0.982	0.973	0.965	0.957
2	2.310	2.275	2.241	2.208	2.176	2.145	2.115	2.085	2.056	2.028	1.973	1.947	1.921	1.896	1.871
3	3.641	3.567	3.496	3.427	3.360	3.295	3.232	3.171	3.112	3.055	2.946	2.894	2.843	2.794	2.747
4	5.105	4.974	4.848	4.727	4.611	4.499	4.392	4.288	4.188	4.093	3.911	3.825	3.741	3.661	3.584
5	6.716	6.506	6.307	6.117	5.935	5.761	5.595	5.437	5.284	5.139	4.867	4.739	4.616	4.498	4.384
6	8.487	8.175	7.880	7.600	7.335	7.083	6.844	6.617	6.401	6.196	5.814	5.636	5.466	5.305	5.150
7	10.436	9.993	9.576	9.185	8.816	8.468	8.140	7.830	7.538	7.261	6.752	6.518	6.295	6.084	5.883
8	12.579	11.972	11.406	10.877	10.382	9.919	9.485	9.078	8.696	8.337	7.683	7.383	7.101	6.835	6.584
9	14.937	14.128	13.379	12.684	12.038	11.439	10.880	10.361	9.875	9.423	8.604	8.234	7.886	7.560	7.254
10	17.531	16.476	15.507	14.614	13.791	13.031	12.329	11.679	11.076	10.518	9.518	9.069	8.650	8.260	7.895
11	20.384	19.034	17.801	16.675	15.644	14.699	13.832	13.035	12.300	11.624	10.423	9.889	9.394	8.935	8.508
12	23.522	21.819	20.276	18.876	17.604	16.447	15.391	14.428	13.546	12.740	11.320	10.695	10.118	9.586	9.095
13	26.975	24.852	22.945	21.227	19.677	18.278	17.010	15.861	14.815	13.866	12.209	11.486	10.823	10.215	9.656
14	30.772	28.156	25.823	23.738	21.870	20.195	18.689	17.333	16.108	15.002	13.090	12.263	11.509	10.821	10.193
15	34.949	31.754	28.926	26.419	24.190	22.205	20.432	18.847	17.425	16.149	13.963	13.026	12.177	11.406	10.706
16	39.544	35.672	32.274	29.282	26.643	24.310	22.241	20.404	18.766	17.306	14.829	13.776	12.827	11.971	11.197
17	44.598	39.940	35.883	32.340	29.237	26.515	24.118	22.004	20.132	18.474	15.686	14.512	13.460	12.516	11.667
18	50.158	44.588	39.776	35.606	31.982	28.825	26.066	23.649	21.523	19.653	16.536	15.235	14.076	13.042	12.116
19	56.274	49.651	43.974	39.094	34.885	31.245	28.087	25.340	22.940	20.842	17.378	15.945	14.676	13.549	12.546
20	63.001	55.164	48.501	42.818	37.955	33.781	30.184	27.078	24.383	22.043	18.212	16.642	15.260	14.039	12.957
21	70.401	61.169	53.384	46.796	41.202	36.437	32.361	28.865	25.853	23.254	19.039	17.327	15.828	14.511	13.350
22	78.541	67.708	58.649	51.045	44.637	39.220	34.620	30.703	27.350	24.477	19.859	18.000	16.381	14.967	13.726
23	87.495	74.831	64.328	55.582	48.270	42.135	36.964	32.592	28.875	25.710	20.671	18.661	16.920	15.406	14.086
24	97.345	82.588	70.451	60.427	52.112	45.189	39.397	34.533	30.428	26.955	21.476	19.310	17.444	15.831	14.430
25	108.179	91.036	77.055	65.601	56.176	48.388	41.921	36.530	32.010	28.212	22.273	19.947	17.954	16.240	14.759
26	120.097	100.238	84.177	71.128	60.475	51.740	44.541	38.582	33.621	29.480	23.064	20.573	18.451	16.635	15.074
27	133.206	110.259	91.858	77.030	65.021	55.251	47.259	40.691	35.262	30.759	23.847	21.188	18.935	17.017	15.375
28	147.627	121.173	100.141	83.332	69.830	58.930	50.080	42.860	36.934	32.051	24.623	21.792	19.405	17.384	15.663
29	163.490	133.059	109.073	90.064	74.916	62.784	53.007	45.090	38.636	33.354	25.392	22.385	19.864	17.739	15.939
30	180.938	146.005	118.706	97.252	80.296	66.821	56.045	47.382	40.370	34.669	26.154	22.967	20.310	18.082	16.202
31	200.132	160.105	129.095	104.930	85.986	71.050	59.198	49.739	42.136	35.996	26.910	23.539	20.744	18.412	16.454
32	221.245	175.460	140.299	113.129	92.004	75.481	62.469	52.161	43.934	37.336	27.658	24.101	21.167	18.731	16.695
33	244.469	192.184	152.381	121.885	98.369	80.123	65.864	54.652	45.766	38.687	28.400	24.653	21.578	19.039	16.926
34	270.016	210.399	165.411	131.236	105.102	84.986	69.387	57.212	47.632	40.051	29.135	25.195	21.979	19.336	17.147
35	298.117	230.236	179.462	141.223	112.223	90.081	73.043	59.844	49.533	41.428	29.864	25.727	22.369	19.622	17.358
36	329.028	251.841	194.616	151.889	119.755	95.418	76.837	62.550	51.468	42.817	30.586	26.250	22.748	19.899	17.559
37	363.031	275.371	210.958	163.279	127.722	101.009	80.775	65.332	53.440	44.219	31.301	26.763	23.118	20.165	17.753
38	400.434	300.998	228.583	175.444	136.148	106.866	84.860	68.191	55.448	45.634	32.010	27.267	23.478	20.423	17.937
39	441.577	328.909	247.589	188.435	145.060	113.003	89.100	71.131	57.493	47.062	32.713	27.763	23.828	20.671	18.114
40	486.834	359.306	268.086	202.309	154.486	119.431	93.500	74.154	59.576	48.502	33.409	28.249	24.169	20.911	18.283
41	536.617	392.413	290.190	217.126	164.456	126.166	98.066	77.261	61.697	49.957	34.099	28.727	24.500	21.142	18.444
42	591.378	428.469	314.029	232.950	175.002	133.222	102.804	80.455	63.858	51.424	34./83	29.196	24.823	21.365	18.599
43	651.615	467.738	339.737	249.849	186.155	140.613	107.721	83.738	66.059	52.905	35.461	29.657	25.138	21.580	18.747
44	717.876	510.507	367.461	267.896	197.952	148.356	112.824	87.114	68.301	54.399	36.132	30.109	25.444	21.788	18.888
45	790.763	557.086	397.360	287.171	210.430	156.468	118.119	90.585	70.584	55.908	36.798	30.554	25.742	21.988	19.024
46	870.938	607.816	429.604	307.755	223.628	164.967	123.614	94.152	72.909	57.430	37.457	30.990	26.032	22.182	19.153
47	959.131	663.066	464.376	329.739	237.587	173.870	129.316	97.820	75.278	58.966	38.111	31.419	26.314	22.368	19.277
48	1056.144	723.241	501.876	353.216	252.351	183.197	135.233	101.591	77.690	60.516	38.759	31.840	26.589	22.548	19.395
49	1162.857	788.777	542.318	378.288	267.967	192.968	141.374	105.467	80.147	62.080	39.400	32.254	26.857	22.722	19.508
50	1280.242	860.153	585.930	405.065	284.483	203.204	147.746	109.452	82.650	63.659	40.036	32.660	27.117	22.890	19.617

Index